BRUNEL
in
SOUTH WALES

THE ENGINEER

I'm a Great Engineer and no end of a swell
And my name it is Isambard Kingdom Brunel
My hat is a stove pipe my seegar puffs steam
And I'm driven along by the quest of a dream.
★★★

In design and invention I can't be surpassed
And each new achievement eclipses the last
When they say I can't do it I show them I can,
And do it twice over, I'm that sort of man.
★★★

By the strength of my bridges great rivers are spanned
And my rails are a network all over the land
My, tunnels dig deep through the rock and the clay
I give orders and nature itself must give way.
★★★

The best and the biggest, the fastest on earth
Are the ships and the engines that I brought to birth
Great Western, Great Britain, Great Eastern, Great Scot!
What have I invented and what have I not.
★★★

For the forgers of iron and hewers of coal
I worked and I planned with strong heart and soul
And from Merthyr to Cardiff I laid down the rails
That brought work and brought wealth to the people of Wales.
★★★

And tall pitheads rose wherever it passed
And houses and streets and a City at last,
For of all who planned wisely and of all who built well
The Greatest was Isambard Kingdom Brunel.
★★★

Harri Webb

Author's comments: I met the late Harri Webb at one of his Poems and Pints meetings in 1978 and he was kind enough to allow me to reproduce this poem in the newsletter of the, now defunct, Brunel Society. He insisted that; '… the spelling <u>seegar</u> is necessary instead of the modern <u>cigar</u>, because the older pronunciation affects the accenting of the word & the ascension of the line. I think it is it is the sort of precision point which an engineer would have appreciated.' See Collected Poems *by Harri Webb (1995), Gomer Press.*

STEPHEN K. JONES

BRUNEL

in

SOUTH WALES

VOLUME III
LINKS WITH LEVIATHANS

To my mother, and in memory of my grandparents.

Like many Cardiff schoolchildren in 1907, my grandfather, Edward Stanley Steadman, received the above medallion to commemorate the visit made by King Edward VII and Queen Alexandra to open the last of the Cardiff Docks: the Queen Alexandra Dock.

Cover illustration: *The* Great Britain *leaving Penarth by Owen Eardley.*

First published 2009

The History Press
The Mill, Brimscombe Port
Stroud, Gloucestershire, GL5 2QG
www.thehistorypress.co.uk

Reprinted 2012

British Library Cataloguing in Publication Data.
A catalogue record for this book is available from the British Library.

ISBN 978 0 7524 4912 8

Typesetting and origination by The History Press
Printed in Great Britain

CONTENTS

FOREWORD

It gives me great pleasure to provide a foreword to Volume III of *Brunel in South Wales* by Stephen K. Jones. In preparing this foreword, I have re-read with interest Volumes I and II, and the former foreword describes the comprehensive coverage provided in the three volumes. In his preface to Volume I, Stephen wrote that 'South Wales has been largely overlooked as an area of interest relating to the works and activities of Isambard Kingdom Brunel (1806–59), yet it was important in terms of engineering landmarks at virtually every stage of Brunel's career. Writing this book is an attempt to redress this and illustrate the achievements and the legacy of the engineer, covering his works against a background of social and industrial history.' The aim stated in the preface to Volume I has now been achieved in full by the author, which is a remarkable achievement bearing in mind the vision and energy of Brunel and the myriad of complex industrial developments which took place in the first half of the nineteenth century in South Wales.

First and foremost, the wide-ranging subject matter covered in these three volumes has been researched in a most thorough manner. Each chapter is supported with detailed references to the sources used in the preparation of the text, and the text is liberally enhanced with suitable quotations and extracts which add to the interest of the reader. Furthermore, the text is illustrated throughout with most appropriate illustrations and photographs, many of which are originals taken by the author. Very often such detailed research can lead to a text which reads like a catalogue; however, Stephen has been able to integrate his research into a most readable and well-illustrated text throughout. This is one of the great strengths of the three volumes.

Brunel's main contribution in South Wales was the development of communications in the widest possible sense. Volume I, *In Trevithick's Tracks*, provides an excellent background for the activities taking place in South Wales between Trevithick's successful attempt to get a steam locomotive to run on rails to Brunel's first commission in South Wales as Engineer to the Taff Vale Railway. Volume II covers a wide range of developing communication links including mail routes, mineral railways and, in particular, the main line across South Wales, with Brunel as the Engineer for the South Wales Railway. This volume, Volume III, deals with maritime communications concentrating on Brunel's magnificent iron steamships. It is of interest to note that communications within Wales and with our neighbours is still a major challenge for the engineers and politicians of the day.

Another major strength of the three volumes is that the author has been able to place Brunel's significant undertakings in an historical context. For example, various contemporary social unrest such as the disturbance around the Castle Inn, Merthyr, in June 1831, which resulted in Richard Lewis (alias Dic Penderyn) to be hanged for his role in the riot, the activities of the industrial agitators known as the 'Scotch Cattle' of Monmouthshire and the Rebecca Riots centred in the west are interwoven in the text. Further afield, the potato famine in Ireland is also introduced as an issue which had an effect on the development of communications across South Wales.

A great deal of interest has been shown in recent times in various aspects of the history of Wales and its people. The language, Welsh literature, religious activities (expansion of chapels, famous ministers of religion, etc.), social deprivation, unrest and political aspirations have been researched and reported in great detail. CADW has extolled the virtues of historic buildings, concentrating on castles, abbeys and monasteries and the properties of landed gentry. The major contribution made by Wales to the Industrial Revolution has not had the attention that it deserves – this is equally applicable in North Wales and mid-Wales, as well as in the south. Whereas a great deal has been done to maintain (at great expense) our architectural inheritance, the same cannot be said regarding our industrial heritage. It is a matter of regret to me that the tourist industry in Wales concentrates its attention on sands and castles and that our industrial past (with the notable exception of Big Pit) has been largely overlooked. These three volumes by Stephen Jones should be made mandatory reading to all people involved in the tourist industry in Wales!

The vast majority of the population of Wales lives in communities developed during the Industrial Revolution. The influx of people into the mineral extraction industries provided the critical mass for a range of social, cultural and sporting activities to thrive. For example, the influx of young people into Cardiff in the last two decades has provided the opportunity for the creation of a number of successful choirs in the same way that the Industrial Revolution provided the opportunity for the creation of choirs, brass bands, and football and rugby teams. When conditions worsened in the industrial communities many sought their future further afield and there is a great deal of interest in the descendents of these people visiting the land of their fathers. Wales should capitalise on this opportunity, and this series by Stephen Jones provides an excellent source of information in order to sell the contribution of Wales to the world.

Finally, I would like to draw attention to a quotation from Volume I: 'The TVR even went to the extent of issuing bibles, kept in station waiting rooms, but clearly expected its workers to think more deeply about the duty they owed to the company in the last of its general regulations (No.284): "In the morning think what thou hast to do; and at night ask thy self what thou hast done".' In view of the extraordinary effort required in producing these three volumes, I would suggest that we know what the answer of Stephen would be to the above instruction by the Taff Vale Railway.

Ben Barr
Emeritus Professor, Cardiff University.

INTRODUCTION

I have been fortunate to have been involved with Stephen Jones since 2005 on expanding the knowledge of Brunel's work in South Wales. There is no doubt that Brunel's considerable involvement in railway and maritime work here has been overshadowed by his works in the South West. It was timely that the culmination of Stephen's studies over many years into the work of Brunel and their crystallisation into this three-volume series was coincident with the 200th anniversary of Brunel's birth in 2006. His first volume *In Trevithick's Tracks* published in December 2005 recording Brunel's original commission for the Taff Vale Railway set the scene for volume two, *Communications and Coal*, published in 2006, describing in detail Brunel's extensive railway works across the whole of South Wales. This third volume, *Links with Leviathans*, completes this study of the influence of Brunel on the industrialisation of Wales, identifying Brunel's later works and his maritime involvement in ports and harbours in South Wales and the steamships he developed to extend his influence overseas. His links [no pun intended] with Samuel Brown and his chains and the Brown Lenox connection with Pontypridd are covered, as is Brunel's extensive work in docks and harbours in South Wales, from Penarth to Neyland.

Stephen's trilogy makes a considerable contribution to the availability of information on Brunel in Wales, bringing a vast amount of dispersed information from a wide range of sources into a concise and readable form, giving life to Brunel's extensive range of activities across the whole geographical area. The considerable amount of information Stephen has amassed over the past forty years is now in the public domain and is retained for posterity. A lifetime of research has been consolidated into three fascinating volumes starting with the development of locomotive-powered railways and culminating in Brunel's maritime endeavours. This final volume continues the examination of Brunel's work in South Wales and its effects on the life and times of its inhabitants, concluding with his aspirations and vision for a transport link from London to Ireland and, ultimately, New York.

There is no doubt that many developments in South Wales resulted from the improved communications engineered by Brunel in his various roles as engineer or consultant to the railway and port organisations. His influence on industrial and social aspects of life in South Wales has now been well documented by this trilogy and his energy, engineering expertise and innovation identified and recorded for all to read. The trilogy concludes with the works of his son, to round off the Brunel contribution to our history and heritage in South Wales.

I am pleased to be able to write this introduction to the final volume of this series and to record my appreciation of the time and effort that has gone into producing this essential reading for all those with an interest not only in Brunel himself but in the development of industrial South Wales and its history.

K.J. Thomas, CBE, C.Eng., B.Sc., FICE, FIHT

Panel Member for Wales
Institution of Civil Engineers Panel for Historical Engineering Works

IMPERIAL AND METRIC DIMENSIONS

As all Brunel's engineering works were executed to imperial measurements these have been retained. The following are the metric equivalents of the Imperial units used:

Length
1 foot = 0.3048 metres
1 yard = 0.9144 metres
1 mile = 1.609 kilometres

Area
1 square inch = 645.2 square millimetres
1 square foot = 0.0929 square metres
1 acre = 0.4047 hectares
1 square mile = 259 hectares

Volume
1 gallon = 4.546 litres
1 million gallons = 4546 cubic metres
1 cubic yard = 0.7646 cubic metres

Mass
1 pound = 0.4536 kilograms
1 ton = 1.016 tonnes

Power
1 horsepower (hp) = 0.7457 kilowatts

Pressure
1 pound force per square inch = 0.06895 bar

ACKNOWLEDGEMENTS

Coal had been a significant factor in the realisation of Brunel's works right from the start of his association with South Wales. *In Trevithick's Tracks* showed that the coal proprietors were needed to provide the critical mass behind the Merthyr ironmasters to make the Taff Vale Railway (TVR) a reality. It was no doubt because of Brunel's involvement with the principals behind the TVR that the Great Western Railway (GWR) placed a large order for Rhondda coal from the Dinas collieries owned by Walter Coffin, the deputy chairman of the TVR. Coal was shipped from Cardiff and converted to coke at the GWR coke ovens in Bristol. The TVR would go on to become one of the most prosperous railways in Great Britain because of the prodigious amounts of coal it carried. This was his only standard-gauge railway and, with the exception of his trunk route, the South Wales Railway (SWR), many of Brunel's broad-gauge railways were promoted and built specifically to exploit the coal measures, despite growing criticism of the unsuitability of the gauge for such purposes. *Communications and Coal* covered these lines, the Vale of Neath Railway (VNR) and the later 'post-crisis' lines, the South Wales Mineral Railway (SWMR), the Llynvi Valley Railway (LVR) and the Ely Valley Railway (EVR). The last named, being one of the first to rebel against its gauge and demands for a uniform gauge, was now being driven by the South Wales coal proprietors. The first two volumes covered these railways, and volume three in the trilogy now focuses on matters maritime. As well as docks and shipping places, there are the associations with his steamships such as the 'great leviathan', the PSS *Great Eastern*, for which Brunel chose Neyland, the terminus of his South Wales Railway, as a home port.

Whilst South Wales' coal has figured as something of a commodity that has shaped the development of railways in the area, the story of the uniqueness of South Wales' coal and its superior quality in steam raising, particularly for steam navigation, will be taken up in this volume in dealing with the PS *Great Western*. Ironically, as this was Brunel's pioneering ship, the SS *Great Britain*, which burnt Welsh coal, would make its last voyage from a coaling dock as a sailing ship.

As well as these connections there is still much to be told, including that of his dock works, shipping facilities and the many maritime connections with his 'leviathans of the sea' to Wales. The *Great Britain* was the first to be called 'leviathan' by the press, and for his final great iron steamship, the PSS *Great Eastern*, would be christened *The Leviathan*. The *Great Eastern* was equipped with chain cable made by Brown Lenox of Pontypridd – the very chain cable that was used as a backdrop to the most famous photograph of the engineer. There were dock works such as Briton Ferry which used his innovative design of a buoyant wrought-iron lock gate. One of his last engineering projects to be realised was a steam railway ferry across the Severn, truly a unique engineering work.

Concluding the story, as far as Brunel is concerned, brings us to his son, his connection with the Penarth Dock, that the *Great Britain* made its last voyage from and his major work in South Wales: Barry Docks. One of my earliest memories of Barry is from the late 1950s when, with my sister Maria, we stayed overnight in Tydvil Street, the home of my Auntie June and Uncle Harry's grandmother. The house was terraced, late Victorian and typical of the many hundreds constructed in Barry following the construction of the docks, that period in which 'volcanic energy … brought Britain through the last century and up to the First World War', to quote the one-time Barry resident, Gwyn Thomas (1913-81). The dock contractor, T.A. Walker, provided the first phase of housing, temporary timber houses on what would become Broad Street, to be followed by speculative houses built by local syndicates to cater for the rapid growth in population. Tydfil Street was one of those developments, and the interior of the house gave me a lasting impression of what it was to live in a Victorian house. From Tydfil Street it was a short walk to Dock View Road, aptly named because it was a vantage point to see, of course, the docks, the railway sidings, the coal hoists alongside the dock and Barry Island beyond. Dock View Road provided an access point to the docks via a subway under the railway tracks, and was no doubt part of the daily route for the original working residents of the area. At the time of these impressions there was little to see in terms of coal export activity; coal trains were in decline. Coal was last shipped here in 1976 and the last coal hoist dismantled in 1981, but the docks continued with imports such as materials for the petro-chemical works that dominated industrial Barry after the end of the Second World War, and bananas from the West Indies. Barry and other docks like Cardiff, and the pioneer in this respect, Swansea, have today been partly developed as marinas, complete with waterside housing.

As in the first two volumes, an attempt is made to place such works in context and to give a contemporary and social insight to the background of such works – all of which form part of the story of *Brunel in South Wales*.

The logos on the back cover of this volume do not signify sponsorship as such but rather a desire to acknowledge the generous support I have received over recent years from the South Wales Institute of Engineers Educational Trust (SWIEET) and the Institution of Civil Engineers, Wales, Cymru, represented by Denys Morgan, Keith Jones and Keith Thomas, Panel for Historical Engineering Works (Insititution of Civil Engineers). For this volume of *Brunel in South Wales* I wish to repeat my thanks to everyone mentioned in volumes I and II for their help and assistance. To Dr Tony Williams, associate editor of *The Dickensian*, Bill Gallagher, William Edwards, Stephen Rowson, Keith Hickman and Simon Hancock, I would like to record my thanks for their constructive suggestions on the text. For help with information, access to archives and illustrations, I would like to thank the following, particularly those who kindly gave me a real link with leviathans, such as Margaret Greenwood whose ancestors worked on the construction of the SS *Great Britain*; Michael Richardson and Hannah Lowery of the University of Bristol (University Library Special Collection); Carol Morgan and Mike Chrimes of the Institution of Civil Engineers (library and archives); Brian George, Sandra Davies, Huw James, Gary Gregor, Ian Pope and John Alsop, Katrina Coopey of Cardiff Central Library; Susan Edwards and the staff of the Glamorgan Record Office; Kim Collis and the staff of the West Glamorgan Archive Service; the Pembrokeshire Record Office; Charlotte Topsfield; Caroline Jacob of the Merthyr Tydfil Reference Library; Audrey David and Gwyn Petty of Porthcawl Museum; Marcus Payne of Penarth Library; Rachael Lovering of the Newport Museum and Art Gallery; Tony Gray of the Ordnance Survey; Kathy Maxwell of the Royal Meteorological Society; the staff of the National Archives, Kew; the staff of the National Maritime Museum, Greenwich; and the staff of the Royal Institution of South Wales and Swansea Museum. The assistance is also greatly appreciated of David and Lorraine Scheeres of Paskeston Hall, Pamela and Thomas Evans of Bute House, Penarth; John Vivian Hughes, Felicity White, Cherry Chronly, Karyn Thompson of ICE, Wales Cymru; Prof. Stuart Cole, Prof. Ben Barr, Peter Keelan and Alison Harvey of Cardiff University (SCOLAR collection); Dr David Wyatt of the Centre for Lifelong Learning; David Claxton, Elaine Davey of the Victorian Society; Roger Worsley, Bryn and Anne Gardner, Martin Barnes, Adrian Andrews, Rodney Cadenne, F.G. Mitchell, Linda Waters, B.H. Wheddon, Alan Thorne, John Pockett, commodore and secretary of Neyland Yacht Club; Paul Reynolds, Gwerfyl Jones, the Falkland Islands Co.; T.E. (Tom) Morgan, OBE, Mary Thomas, Marilyn Holloway; the Welsh Assembly Government, namely Derek Elliott, Jenny Clark, Peter Evans, Adrian Nash, Jason Hibbert, Robert Cone and Robin Shaw; and, as before, extra special thanks go to Owen Eardley, a very accomplished artist who kindly allowed his painting of the '*Great Britain* leaving Penarth' to be used as the cover illustration for this volume, and for his creative input which has been invaluable, in particular in designing new maps and graphics; and to the History Press, particularly Amy Rigg, Emily Locke and James Beer, for their patience and help in completing this volume. As always my greatest debt is to my wife, Siân, for her patience and for assisting me with the research. Similarly, daughters Kathryn and Bethan and granddaughter Chloe cannot be left out! On that note, sincere apologies to anyone I have inadvertently omitted. All errors are regretted and entirely my own.

Stephen K. Jones, 2009

1

GREAT EXPECTATIONS

'I AM MY OWN ENGINEER … AND MY OWN JACK OF ALL TRADES'[1]

The historian of the Great Western Railway, E.T. MacDermot, opened his tribute to Isambard Kingdom Brunel by saying that, 'He was a great engineer of outstanding and original genius…' but completed the sentence by saying, 'too original perhaps in that he was sometimes given to costly experiments; for which the company employing him had to pay dearly.' No doubt MacDermot had the 'atmospheric caper' specifically in mind for those last words which, together with the following remarks on the death of the engineer, can be found in MacDermot's first volume of the *History of the Great Western Railway* in a chapter entitled 'Hard Times':

> On the 15th September [1859] Brunel died, worn out by hard work and worry at the early age of fifty-three. For more than two years his health had been failing, and latterly he had spent much time abroad in Switzerland and Egypt. Though he remained nominally the Great Western Engineer to the time of his death, his position had been practically that of consultant, much as he hated the term… [2]

The reference above to a change in the position Brunel held can be explained by the minutes of the GWR's half-yearly meeting, held at Bristol on 14 August 1857. Of particular interest in the meeting was the change to the terms and conditions that Brunel had enjoyed as the GWR's engineer. The meeting was opened by the Hon. Frederick George Brabazon Ponsonby (GWR chairman, 1857–59)[3] who reported on the expense of providing the narrow-gauge connection to the GWR at Reading but adding that there was now 'no work of magnitude … going on.' A previous chairman, the Rt Hon. Spencer Horatio Walpole (GWR chairman, 1855–56 and 1863) had initiated an enquiry into the company's financial affairs, and the meeting was being conducted in an atmosphere of cost reduction following the publication of this, the Walpole report. Ponsonby assured the meeting that no expenditure beyond what was necessary would be incurred; coal traffic was seen as positive, giving an increase in income from £60,000 to £80,000 a year. Now that all the lines were about to be opened, the directors had considered that the time was appropriate to look at 'remodelling the engineering staff'. Brunel was thanked for the liberal manner in which he had met the committee set up to look into it. The first principle to be laid down was that the heads of the engineering, locomotive and carriage departments should be responsible directly to the Board; 'It necessarily followed that Mr. Brunel's position would not be the same thenceforth. The officers named were to be directly responsible, and Mr. Brunel was to hold the position of consulting engineer.'[4] His salary would be £750 per annum, to include everything, not least commission on contracts, and he would give his advice and assistance when the Board or their officers required it. This would cover all matters connected with public bodies, dealings with other companies and assistance with parliamentary proceedings. The meeting was reminded that if they depended on calling Brunel as a witness in a legal case it would cost as much as his present salary (it was reported that calls of *hear, hear* followed).

The Times reported that a lively discussion followed, particularly when a proposed honourium of £5,000 to the secretary of the GWR, Charles Alexander Saunders (1796–1864), was raised (which itself raised 'murmurs').[5] Saunders was the first secretary to the GWR and his input, along with Charles Russell (1786–1856), the GWR chairman from 1839, and Brunel and Gooch, was crucial in driving forward the company from its early formation and through the broad-gauge period. Saunders' nephew, Frederick George Saunders (1820–1901), had been appointed assistant secretary to the South Wales Railway (SWR) in 1844, taking over as secretary when Captain Nenon Armstrong absconded with £5,000 in 1849. On the resignation of his uncle as GWR secretary in 1863, he was appointed as his replacement.[6] The half-yearly engineers report for this meeting in August 1857 was the last to be signed by Brunel, and the day-to-day work of the GWR was now left to his deputy Thomas Hardy Bertram (1813–89).[7] So, as Ponsonby had indicated, and despite the awarding of honouriums, the GWR was going through a period of belt-tightening and rationalisation, a period of hard times. It is interesting to speculate on MacDermot's choice of chapter title – a reflection of the times, or is there a Dickensian reference being made? There is little to connect the two in any sense of literature, although *Hard Times* was published in 1854, the year that MacDermot opens his chapter.[8] The title of the novel *Great Expectations* by Charles Dickens (1812–70)[9] has, however, been taken as an allusion to engineering achievements and technological progress that held great expectations of success. It goes without saying that there are no similarities in *Great Expectations* and the world of Brunel apart from one very indirect connection, the place of Brunel's forced recuperation on Canvey Island following the fire on the PS *Great Western* steamship in 1838.[10] This was, by all accounts, at the Lobster Smack Inn, used by Dickens as the model for World's End which becomes a hiding place for Pip and Magwitch in *Great Expectations*. The chapter quotation comes from the explanation given

SOUTH WALES RAILWAY.

CONTRACTS FOR WORKS.

NOTICE is hereby given, that TENDERS for the following Works will be received as hereinafter stated :—
Contracts Nos. 11, 12, and 13, Swansea Division, being of the respective lengths of 5 miles 72 chains, 6 miles 16 chains, and 13 miles 32 chains ; and extending from near Kidwelly through Carmarthen to Whitland, county of Carmarthen.
Plans and Specifications may be seen, and printed forms of Tender obtained, at the Engineer's Office, Carmarthen ; and the Tenders must be addressed to the Directors of the South Wales Railway, and delivered at the Railway Office, 449, West Strand, London, on or before twelve o'clock on Wednesday, Feb. 10.
By order. N. ARMSTRONG, Secretary.
London, 11th Jan., 1847.

ONE HUNDRED POUNDS REWARD.—— Whereas NENON ARMSTRONG, commonly called and known as Captain Armstrong, Secretary of the South Wales Railway Company, 449, Strand, has ABSCONDED, having fraudulently negotiated and appropriated to his own use securities for money belonging to the said Company. Whoever will APPREHEND, or give information that will lead to the apprehension and conviction of the said Nenon Armstrong, shall receive £100 Reward. Information to be given at the Police-office, Bow st.; or to Lewis and Nash, Parliament-st., London.

Opposite left: *Brunel's Hungerford Suspension Bridge under construction in the 1840s, soon to sweep away riverside buildings associated with Dickens' youth. It was completed in 1845 but would be replaced a few years later by Charing Cross Railway Bridge. (SKJ collection).*

Left: *South Wales Railway contract advertisement. Advertisement calling for tenders for contracts on 11 July 1847, posted by the secretary, Nenon Armstrong. (The Times, 11 July 1847)*

Left: *Railway respectability turned to robbery just two years later when 'Captain' Armstrong absconded. (The Times, 30 August 1849)*

Right: Brunel *as originally depicted by* The Illustrated London News *with the caption, 'The late Mr. Brunel, Designer of the Great Eastern'. (*Pictorial History of the Great Eastern Steam-Ship, *published in 1859 by W.H. Smith & Son: London)*

to Pip by Wemmick: 'I am my own engineer, and my own carpenter, and my own plumber, and my own gardener, and my own Jack of all Trades.'[11]

For the above, Dickens may have been inspired by descriptions of the novel machinery that had been developed by Brunel for the PSS *Great Eastern*, to describe the various contrivances, including an ingenious fountain, built by Wemmick. However, Wemmick appears to have been inspired by John George Appold (1800–65), an engineer who was present beside Brunel at all of the launching attempts of the *Great Eastern* and who would be on the bridge of the ship when it made its first visit to Milford Haven. Appold was elected as an associate member of ICE in 1850, the first two seconders on his form being Robert Stephenson and Brunel.[12] He was also a prolific inventor who could have been the inspiration for Dickens when describing all the various ingenious contrivances made by Wemmick.[13] In noticing his death, *The Times* referred to his cleverness as an inventor, for example at his house the doors opened as you approached them, 'water came unbidden into the basins' and the shutters closed when the gas lights were lit. Going beyond the novel to the writer reveals experiences common to both Brunel and Dickens, and which no doubt had a bearing on the development of their characters. Both men were born in Portsmouth, to fathers engaged in naval business, although admittedly Marc Brunel's role as the designer of the block-making machinery that was being installed in the Royal Navy dockyard was markedly different to that of a clerk in the navy pay office. The two men, however, would be levelled by the experience of the debtor's prison, and John Dickens' experience in Marshalsea was to be drawn on for graphic effect by his son in *Little Dorrit*.[14] As Rolt remarked in his biography of Brunel, the law on the recovery of debt was such that; '…Marc Brunel was by no means the only eminent man to become familiar with the inside of a debtor's prison, …'[15] The Brunel family had a different reaction to this episode in their family history wherein any discussion of this taboo subject was referred to as 'the Misfortune'. Was there any connection between the writer and the engineer? One possible connection, according to Emmerson, was John Scott Russell (1808–82), who was to take the position of railway editor on Dickens' new morning newspaper, *The Daily News*. [16]

This is about the limit of any obvious similarities between Dickens and Brunel, although both men would go on to become probably the most famous exemplars of their professions, with both their lives being cut short through overwork: Dickens, through the heavy toll imposed by tours of public readings, and Brunel from the sheer volume of works that culminated with the *Great Eastern* and the aggravation of serious illnesses. Stepping back two years from the death of the engineer, to 1857, is useful for reflection, and in reference to MacDermot's comment above on the GWR's engineer, it marks a point of no return. By 1857 Brunel had built the Great Western Railway and numerous associated lines and branches, his greatest iron bridge was in the course of construction, he had two 'Great' ships to his credit along with the hull of his third and final ship, for which description appeared to go beyond superlatives, completed (although the troublesome act of launching the latter ship into the water was still to come). With such a background, and considering the shape of things to come, the year 1857 could be seen as a year of looking forward with expectations of greatness. In South Wales the year was not a particularly auspicious one. The SWR had finally been completed to a western terminus in 1856, and would be complete as a double line on 1 July 1857. The Vale of Neath Railway (VNR), including the Aberdare Valley Railway, was complete by June 1857, and the VNR declared its best ever dividend of 4⅝ per cent in 1857. For the VNR the year 1857 marked a high point but the company's prosperity would now be going downhill. Indeed, inclines were part of the problem due to the high cost of maintaining the permanent way for the heavy mineral traffic, particularly on the Glynneath incline. Light bridge rails were being used on the Glynneath incline which were not up to the task, and even Barlow rails had been considered suitable for the Dare and Amman branches and colliery sidings. Brunel was to design a new type of rail chair and bullhead rail for the VNR, but the costs of these improvements and the depressed state of trade over the winter of 1857–58 led to major changes in the management of the VNR.[17]

Those lines which could be regarded as 'post-crisis' (i.e. lines promoted after the collapse of the 'large' railway mania of 1845–48) included several broad-gauge lines in South Wales, namely the South Wales Mineral Railway (SWMR), the Lynvi Valley Railway (LVR), the Carmarthen & Cardigan Railway (C&CR), the Ely Valley Railway (EVR) and the Bristol & South Wales Union Railway (B&SWUR), none of which would be completed by 1857 or indeed in Brunel's lifetime.[18] The year 1857 was one of the peak years in terms of citations of 'Brunel' in *The Times*, the engineer being mentioned fifty-eight times, a figure matched in 1845 and eclipsed in 1846 with seventy-nine such mentions.[19] Those last two figures reflected the demands on Brunel and indeed the profession of engineering in general, during the economic revival that ushered in the 'large' railway mania in 1845, the year the South Wales Railway (SWR) gained its Act of Parliament.[20] Whilst in 1857 the railway was still developing as a mode of transport, it had long ceased to be a novel form and although there had been some flirtations with alternative forms of propulsion, the pattern that was the steam locomotive railway was now firmly established. Dickens recorded the transition of travelling from the old to the new, but his romantic accounts of the past age of travel tend to identify him with the world of the stagecoach. *Pickwick Papers* and *Nicholas Nickleby* provide such images, which are often referred to as 'Dickensian', particularly when depicted as scenes on Christmas cards. Brunel's brother-in-law, the artist John Callcott Horsley, was well acquainted with such imagery and had included references from Dickens' *A Christmas Carol* in the Christmas greetings card he designed for Henry Cole — the first commercial Christmas card. Horsley had travelled around South Wales with Brunel from as early as November 1834 in connection with making sketches that were intended to be used in the execution of the 'Egyptian' designs for the Clifton Suspension Bridge, and Brunel had relied greatly on coach travel in the overseeing of the construction of his railways and had stayed at numerous coaching inns.[21]

Ironically, most of these coaching inns went out of business when the railway was opened, but some, like the Pyle Inn on the South Wales mail route, where Brunel lodged during the surveying and building of the SWR, survived until 1886. Travellers voted with their feet and, despite the sentimentality of stagecoach travel, it faced swift abandonment whenever there was a choice of transport. Dickens did not experience the steam railway in his youth; he was eighteen by the time the Liverpool & Manchester Railway opened in 1830, and he had closer associations with coastal steamboats. Steamer services operated on the London to Gravesend run and to the estuary towns such as Margate and Southend in the 1820s, and therefore reflected more of the technological images of transport in his early and formative years than the railway did. The railway belongs to Dickens' adult years, as in 1837 when he moved into his first family house in London, at 48 Doughty Street in Bloomsbury, and found himself close to the work going on at Euston.[22] This was the new railway terminus of the first trunk railway; the London & Birmingham. The railway mania years were ones that were graphically recorded by Dickens, who was to capture the spirit of his age in words. One contemporary description describes his writing as if he were 'a special correspondent for posterity'.[23] The novel *Great Expectations* is not one of new social experiences related to industrialisation, but there are many other examples of his work covering such issues. The impact of the railway is picked up at an early stage by Dickens in *Dombey and Son*, with accounts of the building of the London & Birmingham Railway at Tring, Blisworth and Camden Town, and a trip on the railway by Dombey and Major Bagstock. This was the first railway journey fully described by Dickens, a journey that began at Euston by the travellers passing through the now demolished Doric portico, to travel, as was common in those days by the more affluent of society, in Dombey's road carriage secured on a flat railway wagon.[24] They were then taken 'Away with a shriek, and a roar, and a rattle' to Birmingham (Curzon Street Station) to continue by road to Leamington.[25]

Dombey and Son is also the first work by Dickens to feature a railway-related death, Carker, being run down by a South Eastern Railway (SER) train at Paddock Wood. In the same year that the London & Birmingham Railway was opened throughout – from the capital to Birmingham – on 17 September 1838, a section of Brunel's GWR opened between Paddington and Maidenhead with intermediate stops at West Drayton and Slough. Apart from the impressions given of the building of a major railway and a new mode of transport, there are opportunities for a brief discourse on social contrasts – the southerner's impressions of the industrial landscape of the Midlands. Dickens repeated this treatment in 1857 with *The Lazy Tour of Two Idle Apprentices*, describing their journey from London to Carlisle: 'The pastoral country darkened, became coaly, became smoky, became infernal, got better, got worse…'[26] There are many references to the railways of the south-east of England, such as the SER and the London Chatham & Dover Railway, serving his home environs around Rochester.[27] However, in *Our Mutual Friend*, as well as the London & Greenwich Railway, there is a reference to the GWR and Paddington Station, and it also contains a postscript of a train crash that Dickens experienced at first hand. Returning from a journey to France on 9 June 1865, Dickens had taken the cross-Channel ferry at Boulogne, resuming his train journey at Folkestone. The train was bound for Charing Cross but disaster struck at Staplehurst where track work was being undertaken, and, despite an attempt to slow it down, the train was derailed and crashed with seven carriages plunging into the River Beult.[28] Of the 110 passengers onboard, ten were killed and fourteen badly injured, with Dickens tending to the injured and dying. Naturally he would describe his experiences of such an ordeal and record the anger he felt at the gross incompetence that had caused the accident. Dickens was accompanied on this journey by the actress Ellen Ternan, with whom he had been conducting an affair since their first meeting in 1857. Dickens was to undertake numerous 'clandestine' journeys on the GWR in order to visit her at Slough.[29]

Dickens first tried this new mode of travelling in 1838 and continued to use it for the rest of his life, although the shock he suffered in 1865 put additional nervous strain on him, and no doubt was part of the strain on his health that contributed to his death in 1870.

Whilst a number of Victorian novels, particularly those of Dickens already mentioned, had a degree of industrial content, readers seeking the genre of industrial biography would be enlightened by a writer born in the same year as Dickens; Samuel Smiles (1812–1904). Indeed, Smiles was to dominate this field after a number of career changes from surgeon to newspaper editor, and he also had first-hand experience of railway development, working as secretary to the Leeds & Thirsk Railway in 1845, followed by a similar post at the SER in 1854. Today Smiles is best remembered for his descriptive *Self-Help* of 1859, but he was to cover the lives of such engineering luminaries as James Watt, Thomas Telford and both George and Robert Stephenson. *George Stephenson* was in fact the first of his works on the lives of the engineers, published in 1857. Brunel was, for some reason, a remarkable omission from a long list of engineering biographies.[30] Perhaps Smiles felt he could not fit Brunel into his standard template of the self-made or modest engineer, as he had done, respectively, with George and Robert Stephenson. Smiles' descriptive accounts of moral conduct would result in a number of works with titles such as *Character*, *Thrift* and *Duty* in the years up to 1880. Smiles's Victorian engineers represented the moral rewards of hard work and thrift – a view of financial failure, with the consequences of debtor's prison, having already been sketched by Dickens in *Little Dorrit* (1855–57). Mr Merdle in *Little Dorrit* is not based on a character like George Hudson (1800–71), the 'Railway King', who had been brought down with the collapse of the 'large' railway mania, but a certain Tipperary banker and MP for Sligo, John Sadlier. There is a reference to '…a certain Irish bank' in the preface of *Little Dorrit*, a bank in which Sadlier had been involved in the controversial dealings of railway shares. The subsequent failure of the bank led to Sadlier's suicide.[31]

A more graphic description of the 'unacceptable face of capitalism' was Anthony Trollope's fictional story of the financing of railway undertakings in *The Way We Live Now* in 1875. This work chartered the rise and fall of the railway speculator and charlatan-financier Auguste Melmotte, a character based on the real life Jewish 'Baron' Grant (originally Albert Gottheimer). Trollope may also have drawn some inspiration from the debacle of the Tipperary Bank as Melmotte commits suicide by the same method (poison) as Sadlier. There was plenty of inspiration for what might be called 'creative accounting' from revelations following the downfall of Hudson who, like Sadlier (and indeed the fictional Melmotte), were Members of Parliament. Hudson was also used as the model for Disraeli's character, Mr Vigo, a tailor who comes from Yorkshire, in his last novel, *Endymion*. Mr Vigo becomes the chairman of the Great Cloudland Railway and gets involved in railway construction in Yorkshire and Lancashire.[32]

Robert Stephenson (1803–59) in later life. (SKJ collection)

*George Hudson MP, 'The Railway King' (*The Illustrated London News*, 6 September 1845)*

'*Mr. Scott Russell, Builder of the Great Eastern'.* (Pictorial History of the Great Eastern Steam-Ship, *published in 1859 by W.H. Smith & Son: London)*

Like Merdle, Melmette can be seen as an amalgam of all fraudulent financiers based on any number of real life examples. [33] Robert Stephenson (1803-59) was associated with Hudson on a number of railway schemes and somewhat naturally favoured railways over other forms of transport such as ship canals. This was viewed by some as biased particularly when, as the only Englishman on the Suez Canal Survey Committee, he pronounced against it. Using his position as an MP in the House of Commons he also stated that no English capitalists would benefit from it, although others might become richer. The moving force behind the Suez Canal, Ferdinand de Lessops (1805-1894) took offence to what he thought was a slur of 'Melmotte' proportions to his character. He promptly, on 27 July 1857, challenged Stephenson to a duel. [34]

Hudson had little professional contact with Brunel but there was the occasion in 1844 when both were in opposite camps of competing railway schemes for the completion of the London to Edinburgh route, Brunel being appointed engineer for the Northumberland Railway proposal against the Stephenson-engineered Newcastle & Berwick Railway. The fact that Brunel's line was to be engineered for the atmospheric system of propulsion was not well received by the Board of Trade, and this, along with other factors concerned with the actual route, would lead to the approval of Hudson's line. [35] The atmospheric system of propulsion suffered a setback on this occasion but Brunel was still intending to use it on the South Devon Railway (SDR) and possibly on the final section of the SWR, planned to serve the proposed terminal port at Abermawr. [36]

However, the problems associated with operating the SDR by atmospheric propulsion with the technology available eventually proved to be insurmountable and it was abandoned in September 1848. [37] Ranking as Brunel's greatest engineering failure, the atmospheric system had promised 'great expectations' but even Brunel could not make it deliver. Just a year or so after the rejection of the Northumberland Railway Bill, during the time that atmospheric pipes were being laid on the SDR, Hudson had the opportunity to give his opinion to Prince Albert on atmospheric railways. They were, in his own words, an 'umbug', but the British public would soon discover that Hudson was the greatest humbug of all. Indeed, in May 1847 Dickens was to refer to him as '…the Giant Humbug of this time…'. [38] Hudson had moved in the best of circles but now faced enquiry after enquiry for financial impropriety and ultimately imprisonment for debt. Prior to the collapse of his empire he was someone who the Merthyr ironmasters relied upon in the business of supplying rails. As well as Merthyr, the Crawshay family had manufacturing interests at Gateshead and would win the contract to build Stephenson's High Level Bridge at Newcastle. [39] When trying to avoid charges of fraud relating to the sale of rails, Hudson had claimed that William Crawshay II (1788–1867) attempted to bribe him to reject a tender from Alderman William Thompson of Penydarren (d.1854) to supply rails to one of his companies. If he had accepted the bribe, Crawshay would have supplied the rails at £14 a ton (Thompson tendered at £12 a ton) and Hudson would have pocketed £50,000. [40] It was Thompson who had given Brunel the go-ahead for the survey that would lead to the building of the Taff Vale Railway (TVR).

Going back to this and one of Brunel's earliest visits to South Wales, in connection with survey work for the TVR, the engineer visited Dowlais for a meeting with Sir John Guest (1785–1852) on 12 October 1834. At the meeting, Lady Charlotte Guest (1812–95) referred to Brunel in her diary as '…Mr. Brunel of the Thames Tunnel'. The association of the Brunels, father and son, and the image of the tunnel, rippled out for many years to come, Brunel being associated with it in the early years of his independent career, even though he only played a supporting role in what was his father's last major project. An interesting observation was made during a tour of southern Italy in 1847 by Edward Lear (1812–88), who was born in the same year as Dickens and whose father had also been confined to a debtors' prison. [41] Lear was an accomplished artist, writer and illustrator of books of travel but is remembered more for his 'nonsense' verses. A commissioner of Lear's oil paintings was Henry Austin Bruce, later 1st Baron Aberdare (1815–95), who would be critical of Brunel's progress in completing the VNR.[42] While travelling through Calabria, the 'toe' of Italy, Lear stayed at the monastery of Santa Maria di Polsi where he was asked by the Padre Superior why he had come to such a solitary place, as 'No foreigner had ever done so before!' He, however, had a view of England which he shared with Lear, telling him that it was a very small place, about a third the size of Rome, but thickly inhabited by a people that were a sort of Christian but:

> Their priests, and even their bishops, marry, which is incomprehensible, and most ridiculous. The whole place is divided into two equal parts by an arm of the sea, under which there is a great tunnel, so that it is all like one piece of dry land. Ah – che celebre tunnel![43]

That the 'celebre tunnel' should have such a high profile in a rural part of southern Italy at the time Lear was travelling (1847), which was some nineteen years after work on the Thames Tunnel had first begun and four years after it was finally completed, is an indicator of one of the monuments that the elder Brunel had given to the world.[44] The younger Brunel would also give the world incredible monuments, one of which, the *Great Eastern*, Lear would not only see for himself but would have the opportunity to sketch on 9 September 1859.[45] Dickens had something to say on the *Great Eastern*; his first personal observation appears to have occurred some time before Brunel's death but only put down on paper the day after Brunel's death, according to a letter written to Wilkie Collins on 16 September 1859. In this he states that he went right up to London Bridge by the boat that day, 'on purpose that I might pass her'. Dickens had very strong views: 'I thought her the ugliest and most un-ship like thing these eyes ever beheld.'[46] It was Dickens' Special Correspondent, John Hollingshead, who was to witness at first hand the early, key moments of the *Great Eastern*, such as the launch in 1857 and the tragic trial voyage of 1859. His observations are not as unkind to the leviathan itself, possibly the sheer scale and concept of the ship could only be fully appreciated for their potential by a new breed of storyteller, the science fiction writer. The man regarded as the originator of this form, 'the expander of horizons', Jules Verne, also experienced at first hand a voyage on the *Great Eastern*. This was some eight years after Hollingshead's reports of 1859, when Verne recorded his first impressions of the ship in *A Floating City*:

> One would have taken her for a small island, hardly discernable in the mist. She appeared with her bows towards us, having swung round with the tide; but soon the tender altered her course, and the whole length of the steamship was presented to our view; she seemed what in fact she was – enormous![47]

Brunel compared the construction of the great iron ship to the design of large wrought-iron bridge, saying of the latter, that if it had been built, it could have been called the 'leviathan' of bridges; a single span crossing of the River Severn. This was based on his design being worked out for the railway crossing of the River Tamar, a bridge that would be opened as the Royal Albert Bridge, and on 1 September 1857 the first truss was floated out. Nearer home, the Crumlin Viaduct was opened, four years after construction began, on Whit Monday 1 June 1857.[48] The viaduct was constructed as part of the Taff Vale Extension and formed a link between the Newport, Abergavenny & Hereford Railway (NA&HR) and the TVR. As a cross-valley railway, the line had to overcome some major natural obstacles such as the Ebbw valley at Crumlin, where the viaduct (actually, two separate viaducts) and what was Britain's tallest ever railway structure, would stand 200ft above the valley floor. The choice of using the Warren truss, patented by James Warren (1802–70) in 1848,[49] was arrived at after consideration

A Welsh achievement for 1857; the Crumlin Viaduct, photographed by Francis Bedford in the 1860s. (Stephen Rowson collection)

of the option of a masonry viaduct was ruled out because such a viaduct able to withstand the buffeting of winds was considered too costly to build. There is some suggestion that a Brunel-style timber viaduct was considered, but the NA&HR's engineer, Charles Liddell (*c*.1813–94), decided on a design by Thomas William Kennard (1825–93) which used prefabricated riveted wrought-iron members joined together with wrought-iron pins to make up the Warren truss girders.[50] The cast iron came from Kennard's foundry in Falkirk, with the wrought iron supplied by the Blaenavon Iron Co., and was prefabricated into girders at a purpose-built works at Crumlin. In the same way that Brunel's Wye Bridge at Chepstow led to the setting up of what was to become a permanent bridge works on site, so the Crumlin Works was established which exported girderwork all over the world.[51] There was still no sign of a resumption of work on Brunel's first independent commission, the Clifton Suspension Bridge, but a new railway into London was being planned which would have implications here.[52]

Whilst his fellow engineer Robert Stephenson took to sailing his yacht as a form of relaxation and escape from frenetic work schedules, Brunel sought an escape in another direction, in a move he had been planning for some years. He took the first step in this direction in 1847 when, at the height of the 'railway madness' of those years, he purchased the first plot of land for his country retreat, or sanctuary, away from the pressures of his engineering workload, at Watcombe. The house itself was destined never to be realised, at least not to Brunel's plans; however, he would pay many visits to Watcombe, or Watcombe Park as it became known, particularly in 1857. He would visit with his family to inspect the development of the park, which Brunel built up over the years by acquiring parcels of land and directing the landscaping, laying out and planting of the park with trees and shrubs, many of which were exotic species. A schoolboy staying in the area during the Easter of 1858, Arthur James, would record his memories of a visit to Watcombe when Brunel was in 'residence'.[53] James, who was to marry Brunel's daughter, Florence, saw his future wife's father for the first and only time, an impression of whom he would remember later as '…a little business-looking man in seedy dress with a foot-rule in his hand.'[54] Playing the part of Brunel's 'assistant or resident engineer' there for almost ten years was Alexander Forsyth, his head gardener at Watcombe, who like many of the leading gardeners of that time was Scottish. Brunel's son Isambard reflected on this time: 'There can be little doubt that the happiest hours of his life were spent in walking about in the gardens with his wife and children, and discussing the condition and prospect of his favourite trees.'[55]

Brunel enjoyed the work of altering the landscape, with new drives and vistas, so that he could enjoy nature, and the following quotation from Robert Louis Stevenson (1850–94), an account of 'An Old Scotch Gardener', seems appropriate in this respect: 'Annihilating all that's made – To a green thought in a green shade.'[56] Another of the great Victorian storytellers, Stevenson was an engineer's son, and one who originally intended to go into engineer-

Left and above: *The Blue Ribbon trophy is awarded to the ship that makes the fastest crossing of the Atlantic. The* Great Western *is commemorated on the trophy. (Courtesy of the US Department of Commerce, United States Embassy, London)*

ing and started his university education to this end. Despite abandoning engineering for law, Stevenson, the only son of Thomas Stevenson (1818–87) and grandson of the famous Robert Stevenson of Bell Rock lighthouse fame and the founder of this engineering dynasty, is best remembered for his literary work. A chapter on his father, 'Thomas Stevenson: Civil Engineer', appeared in a slim volume entitled *Memories and Portraits*, whilst a fuller account of the engineering dynasty would be published posthumously as *Records of a Family of Engineers* in 1896.[57]

The year 1857 did bring one phase of Brunel's achievements to a close, that of his first venture into steam navigation, for in the early part of that year his first steamship, the wooden paddle steamer *Great Western*, was being towed upriver from Israel Marks' Greenwich wharf to her final berth. Launched twenty years earlier in 1837 from the Bristol yard of Patterson and Mercer, the *Great Western* was to complete forty-five transatlantic voyages for her original owner, the Great Western Steamship Co. The company were forced to sell the *Great Western* in order to fund the cost of salvaging the *Great Britain* after she ran aground in Dundrum Bay. Consequently, in 1847 the *Great Western* was bought by the Royal Mail Steam Packet Co. (RMSPC) for its service between Britain and the West Indies, and subsequently to South America. From February 1856 she was engaged as a Crimean War troop transporter, being requisitioned by the Admiralty and designated as 'No.6 Transport'. These were to be her last revenue earning voyages, the last leg bringing back men and equipment of the Royal Artillery from Scutari and Balaklava to Portsmouth in June 1856.[58]

Following her discharge from war service, the directors of the RMSPC decided that they could not justify the cost of the repairs necessary to put her back into mail service, and the *Great Western* was sold as a job lot with another RMSPC steamer, *Severn*.[59] Israel Marks ran a brass foundry and iron merchants, and was also in the scrap business, removing all the metal parts, such as the engines and boilers, that he could from the ship. At low tide he stripped the copper sheathing from the hull and removed the ship's side valves. When he was finished the sad remains of the once great ship were sold to Henry Castle, whose breaker's yard was at Millbank, Vauxhall. In order to get there, the masts and part of her upper deck had to be removed to allow her to pass under the Thames bridges on the way. The *Great Western* was again towed, this time to her final berth at Castles' Wharf at Vauxhall, a journey that took her past Millwall on the Isle of Dogs and the towering spectacle of Brunel's new leviathan of the seas; the *Great Eastern*. For a brief instance the first and the last of Brunel's ships were in hailing distance, although the 236ft-long *Great Western* was, in common with every other ship afloat at the time, dwarfed by the 693ft-long iron hull under construction. What they had in common, apart from Brunel, was that they were both 'leviathans' – the largest ships in the world at the time of their respective launches.[60] Despite the almost unrecognisable condition of the ship in 1857, her place in history as the first ocean-going steamship was assured, and her last voyage did not go entirely unnoticed: 'Among those who went there to take a farewell of her before she finally disappeared was Mr. Brunel; thus he saw the last of his famous ship.'[61]

CHAPTER 1 NOTES

1 The quotation comes from a conversation between Wemmick and Pip when Pip visits his wooden house complete with various contrivances, including an ingenious water cooler and a fountain. Here Wemmick tells Pip, in acknowledgment of his compliments; 'I am my own engineer, and my own carpenter, and my own plumber, and my own gardener, and my own Jack of all Trades,' p.183 of Collins edition (1961) of *Great Expectations* (first published in 1861).

2 MacDermot, E.T., revised by Clinker, C.R., p.222 (1964), *History of the Great Western Railway, Vol.1* (London: Ian Allen). See his observations on the term 'Consulting Engineer' in Vol.1, p.121.

3 Previously Ponsonby had been the chairman of Pilbrow's Atmospheric Railway & Canal Propulsion Co., Pilbrow's patent sought to avoid the problems of a fixed connection between the train and the piston in the atmospheric tube.

4 *The Times*, 15 August 1857.

5 Calls from the floor included the proposal to set up a committee of investigation into the company. This was then amended so as not to suspend the dividend of ½ per cent, but to refer everything else to this committee.

6 He is one of the few examples of individuals associated with the railway to be remembered, contemporaneously, with a Cardiff street name; Saunders Road being the access road to the SWR, later GWR, Cardiff Station, from St Mary Street to Wood Street, an access also known at the St Mary Street intersection as 'Great Western Approach'. An examination of this legacy will be covered in Chapter 10

7 MacDermot, E.T., revised by Clinker, C.R. (1964), p.222.

8 MacDermot, E.T., revised by Clinker, C.R. (1964), Chapter IX, 'Hard Times' (1854–63).

9 *Great Expectations* was published initially in serial form in 1860–61, and so appeared after Brunel's death.

10 On 31 March 1838 the PS *Great Western* caught fire on her way back from London to Bristol and was forced to 'beach' on a sandbank off Canvey Island. Brunel fell some 18–20ft off a ladder into the boiler room while trying to help and, seriously hurt, was taken ashore to Canvey Island.

11 Dickens, Charles (1961 originally pub. 1861), p.183.

12 Appold's cable brake was used on the *Great Eastern* for the laying of the telegraph cable, see Chapter 8 His centrifugal pump, used to demonstrate a series of cascades or waterfalls, was a great attraction at the 1851 Great Exhibition.

13 ICE membership records, 9 April 1850 (elected 7 May).

14 Marc Brunel was imprisoned in the nearby King's Bench Prison in London.

15 Rolt, L.T.C. (1959), p.15, *Isambard Kingdom Brunel* (Longmans, Green & Co.: London).

16 Emmerson, George S. (1977), p31 *John Scott Russell, A Great Victorian Engineer and Naval Architect*, John Murray, London. However, Dr Tony Williams is not aware of John Scott Russell's connection

and points out that in John Drew's book, *Dickens and the Journalist* (2003) a William Scott Russell is mentioned as the railway editor (p820.

17 The Directors of the VNR reported in 1858 that Brunel's role was to be one of 'Consulting Engineer'. For Brunel's design of a special double-headed rail chair see Vol.II, p.191 and 196, also illustration on p.190.

18 See Jones, Stephen K. (2006), p.77, for an explanation of what is meant by 'post-crisis'.

19 *The Times* citations for Brunel, searched using the online facility, came up with the following: 1825 (7), 1826 (2), 1827 (16), 1828 (10), 1829 (4), 1830 (5), 1831 (2), 1832 (5), 1833 (9), 1834 (1), 1835 (10), 1836 (32), 1837 (7), 1839 (5), 1840 (4), 1841 (13), 1842 (14), 1843 (16), 1844 (52), 1845 (58), 1846 (79), 1847 (36), 1848 (41), 1849 (19), 1850 (51), 1851 (46), 1852 (33), 1853 (27), 1854 (24), 1855 (16), 1856 (39), 1857 (58), 1858 (16), 1859 (44), 1860 (45), 1861 (15), 1862 (19), 1863 (10), 1864 (21), 1865 (3). Obviously some of these citations, particularly around the mid-1820s, would include references to his father, Marc Isambard Brunel, the first of which appears to have been in 1817, and from then until 1824 there were thirty citations.

20 For an explanation of the 'little' and the 'large' railway manias, see Vol.I, p.103.

21 For Horsley, see Vol.1 pp.81–82, and for an example of a South Wales coaching inn see Vol.II p.135.

22 Williams, Tony (2006), *Dickens and 'The Moving Age'*.

23 Williams, Tony (2006), *Dickens and 'The Moving Age'*. The 'special correspondent' is Walter Bagehot reviewing Dickens' work to date in the National Review No.7, October 1858.

24 A view of a loading bay for horse-drawn carriages can be seen in illustrations of Bridgend in Vol.II; colour plate 24, and an early photograph on p.132.

25 House, Humphrey (1942), pp.139–141, *The Dickens World* (Oxford University Press: London).

26 House, Humphrey (1942), p.142. *The Lazy Tour of Two Idle Apprentices*, Chapter 1, Household Words, Vol.XVI, 3 October 1857.

27 House, Humphrey (1942), pp.19–20. The expression 'Dickens Land' is used by the author for the use he made of the Rochester-Chatham district as models for his fictional places. Other works with railway themes include Mugby Junction, a collection of eight short stories of which three are centred on Rugby 31 Station, and three are on other railway subjects including the first railway ghost story, *The Signalman*.

28 Williams, Tony, *Dickens and 'The Moving Age'*.

29 Ottley, George (1965 second edition 1983), *A Bibliography of British Railway History* (Science Museum/National Railway Museum: London). Ref. 7472 for Alymer, F,. (1959), *Dickens Incognito* (Hart-Davis: London).

30 Except a brief mention in a review by Smiles on Richard Beamish's biography of Marc Isambard Brunel.

31 Dickens also mentions the collapse of the Royal British Bank, which had invested heavily in Welsh gold mines, believing Wales to be the next 'California'.

32 Bailey, Brian (1995), pp.160–163, *George Hudson: The Rise and Fall of the Railway King* (Alan Sutton Publishing Ltd: Stroud). The author also refers to similarities with Hudson and the characters Edward Pondervo in *Tono Bungay* (by H.G. Wells) and Karsten Bernick in *The Pillars of Society* (by Ibsen).

33 Examples that continue to live amongst us now, such as the multi-billion dollar fraudster Bernard Madoff, sentenced on 29 June 2009 to 150 years in the USA for fleecing investors out of an estimated $65 billion (£45 billion).

34 Stephenson would write to de Lessops to say that he had been, to use a modern politician's line, misquoted, and that none of his remarks could be construed as alluding to him, see p.84, pp.158-160, Charles Beatty (1956), *Ferdinand de Lessops*, Eyre & Spottiswoode, London.

35 It is not intended to cover the atmospheric episode in any detail here, but further details of the Northumberland Railway can be found in Chapter 2.

36 Brinton, Piet and Worsley, Roger (1987), pp.123–128, *Open Secrets: Explorations in South Wales* (Gomer Press: Llandysul).

37 Professor Jack Simmons in his book *The Victorian Railway* (1991), refers to the atmospheric episode as

a case of '…demand defeated by an insufficient technology' (p.73). See also Buchanan, R.A. (2002), pp.106–112, *Brunel, The Life and Times of Isambard Kingdom Brunel* (Hambledon & London: London).

38 Bailey, Brian (1995), p.65 and 79.

39 If Brunel's Northumberland Railway had been successful it would have also crossed the Tyne.

40 Bailey, Brian (1995), p.105. See also Vol.I, pp.97–98 for an account of a London dinner party hosted by Thompson and attended by the Hudsons and the Guests.

41 Published in 1852 as *Journals of a Landscape Painter in Southern Calabria*, see the edition introduced by Quennell, Peter (1964), *Edward Lear in Southern Italy* (William Kimber: London). Lear, having the Earl of Derby as a patron, had travelled there to paint landscapes.

42 See Vol.II, p.197. These were for his house at Duffryn near Aberdare, one of which now hangs in the National Museum and Galleries of Wales in Cardiff.

43 Quennell, Peter (1964), p.76. Calabria was then part of the Kingdom of Naples, the King of Naples would be one of the casualties of the following year of 1848, 'the year of revolutions'.

44 See Chapter 1, Vol.I.

45 Levi, Peter (1996), p.170, *Edward Lear: A Biography* (Papermac: London).

46 The description would imply that Dickens saw her in her unfinished state awaiting fitting-out.

47 Verne, Jules (originally published in 1871, Fredonia Books edition, 2002), pp.4–5, *A Floating City* (Fredonia Books: Amsterdam).

48 The bridge was closed in 1962 but had been scheduled as being of architectural and historical interest by the Ministry of Housing and Local Government. However, by 1964 this decision had been overruled by British Railways and demolition began in 1966. Whilst demolition was in progress, Universal Studios used the viaduct to shoot scenes for *Arabesque*, a tongue-in-cheek spy thriller starring Gregory Peck and Sophia Loren.

49 It was a joint patent with Willoughby Theobald Monzani, Proc. Inst. C. E., Vol.XII, 1852–53.

50 Smith, Martin (1994), pp.64–66, *British Railway Bridges and Viaducts* (Ian Allan Publishing: Shepperton).

51 Berridge, P.S.A. (1969), pp.156–7, The *Girder Bridge After Brunel and Others* (Robert Maxwell: London). Crumlin was also the subject of one of the first papers presented to the South Wales Institute of Engineers; *On the Crumlin Viaduct and on Wrought Iron Beams and Girders*, by Henry N. Maynard, Transactions, Vol.II, pp.41–54.

52 This necessitated the removal of his Hungerford Suspension Bridge across the Thames, and the demolition and salvage of suspension chains and other material from the bridge would lead, finally, to the completion of Clifton.

53 Brunel rented a number of houses in the area, the longest tenancy being at Watcombe Villa, now Watcombe Lodge, from August 1849 to March 1857. After this he took up residence at Portland Villa, now part of the Maidencombe House Hotel.

54 Tudor, Geoffrey and Hilliard, Helen (2007), p.109, *Brunel's Hidden Kingdom* (Creative Media Publishing: Paignton).

55 Tudor, Geoffrey and Hilliard, Helen (2007), p.102.

56 Stevenson, Robert Louis (1920), p.55.

57 Stevenson, Robert Louis (1920), *Memories and Portraits* (Chatto and Windus: London).

58 Griffiths, Denis (1985), pp.132–3, *Brunel's 'Great Western'* (Patrick Stephens: Wellingborough). Griffiths mentions that the SS *Great Britain*, also requisitioned as a troopship, was in the same waters at the same time.

59 Griffiths, Denis (1985), pp.133.

60 The *Great Western* would be exceeded by the *Great Britain* for that title, but the *Great Eastern* would remain the largest ship in the world, even after her demise, and not be exceeded in all dimensions until the launch of the *Mauretania* in 1907.

61 Brunel, Isambard (1870, reprinted 1971), p.244, *The Life of Isambard Kingdom Brunel* (Longmans, Green & Co.: London) (reprinted David & Charles: Newton Abbot).

2

STANDING TOGETHER

'I ASKED MR LENOX TO STAND WITH ME, BUT HE WOULD NOT SO I ALONE AM HUNG IN CHAINS'[1]

Through their individual talents and achievements, Brunel and Robert Stephenson raised the profile of the engineer to an unprecented level and, by the synergy of 'standing together', they greatly enhanced public awareness of their profession. Whilst Brunel's professional contact with Stephenson, and indeed other contemporaries, was wide ranging, the instances of working together are few and far between. One such example in Wales of possible collaboration with another engineer was that with Charles Blacker Vignoles (1793–1875) on the proposed broad-gauge line to a new Irish packet port at Porthdinllaen.[2] There are examples of his work with engineers and advisers, men like Captain Christopher Claxton (1790-1868), Thomas Guppy (1797–1882), Thomas Sopwith (1803–79) and, of course, John Scott Russell (1808–82). The professional rivalry between Brunel and Stephenson is evident from the beginning of their engineering careers, from the time they came together as the respective engineers of the Great Western and the London & Birmingham in the negotiations for access to Euston Station in 1836 to when they faced each other at Cardiff over the Ely proposals.[3] Frequently they were found giving evidence on opposite sides concerning the various railway schemes going through parlimentary enquiries and committees. But, despite their professional allegencies, they remained on excellent terms, friends who supported each other's engineering projects with their presence and practical advice. This was evident at occasions such as the floating out of the first iron tube of the Britannia Bridge when Stephenson invited Brunel and Joseph Locke (1805–60). These three, the 'great trimuvarate of engineers' as the Victorian press later referred to them, gathered on the top of the great tube along with the senior members of Stephenson's team: Charles Heard Wild (c.1819–57) and Edwin Clark (1814–98), along with Captain Claxton, Sir Francis Head and the contractor, Christopher John Mare (1815–99). Brunel reciprocated and invited Stephenson to the floating out of the Chepstow trusses and the launch of the *Great Eastern*.[4] In 1846, before the first of these high-profile gatherings, Brunel had remarked that: 'It is very delightful in the midst of our incessant personal professional contests carried to the extreme limit of fair opposition, to meet him on a perfectly friendly footing and discuss engineering points.'[5]

Locke, who only just outlived his fellow engineers, and would act as one of the six pall-bearers at Stephenson's funeral, referred to the death of his fellow engineers in his presidential address, commenting that the Insitution of Civil Engineers (ICE) had every reason to be proud of its association with such names as Brunel and Stephenson. The title of trimuvarate was largely a posthumous one based on the legacy of their individual careers, as there were few instances of collaborative work apart from serving together on the building committee of the Great Exhibition of 1851, or on numerous committees set up to look at engineering matters or public health issues. The trimuvarate, however, did join together to lobby issues such as cam-

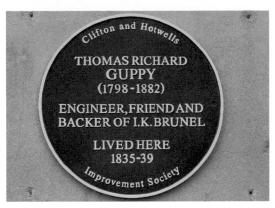

Above and left: *Thomas Richard Guppy (1798–1882) and the plaque on 10 Berkeley Square, Bristol. Charles Richardson (1814–96) also lived here. Guppy would also be involved with the Cwmavon Works near Port Talbot. (Portrait courtesy of University of Bristol special collections, plaque SKJ photograph)*

Far left: *Thomas Sopwith (1803–79), the Newcastle-born engineer who would assist Brunel with the Northumberland Railway surveys. (Courtesy of the Royal Metrological Society)*

Left: *Joseph Locke (1805–60), one of the great triumvirate of engineers. (From the 'Conference of Engineers at the Menai Straits...' by John Lucas, courtesy of the Institution of Civil Engineers)*

paigning for the Ordnance Survey to complete the national mapping scale at 1in to the mile in 1853, appealing for a 20in scale in urban areas.[6] The major issues that Brunel and Stephenson championed, and which often put them on opposing sides, were those related to atmospheric propulsion versus steam traction and railway routes based on the narrow and broad gauges. Brunel, however, refrained from opportunities to represent opposing interests, such as criticising the design of Stephenson's trussed compound girder in the inquest held on the Dee Bridge collapse in 1847. He would write that he had 'such a horror of the dog-eat-dog system of warfare which has grown up amongst professional people from Parliamentary contests...'

and he did not want to attack a 'professional brother', particularly Robert Stephenson.[7] But before he created his 'broad gauge' as the *Great Western* engineer, Brunel had put himself forward as engineer to proposed railways in the Midlands and the north of England, very much the territory of his 'professional brother'. In January 1830 he was unsuccessful regarding the Newcastle & Carlisle Railway but in the November of the same year he met with the promoters of the proposed Bristol & Birmingham Railway, writing that 'Something may at last turn up of this'. He was to be disappointed again as the Bristol & Birmingham scheme failed to get sufficient support, even though he carried out some preliminary work and surveyed a possible route, to the east of the line adopted by the later Birmingham & Gloucester Railway.[8]

A successful 'northern' commission in this period was for a new dock at Monkwearmouth, near Sunderland, which he secured at the end of 1831, but Parliament rejected the first proposals and when eventually an Act was passed it was for a scaled-down scheme. Then came the great opportunity of the Great Western Railway (GWR), and his interest in railways was largely confined to what was regarded as 'GWR territory' in this country. That is until the mid-1840s, with one such projected railway scheme being attributed to him at the beginning and then to almost every engineer of standing, excepting Robert Stephenson.[9] Ignoring this, the line that would bring Brunel and Stephenson together in direct conflict was the proposed Northumberland Railway of 1844 to which Brunel had agreed to become engineer at the request of Lord Howick.[10] Howick persuaded Brunel not only to enter 'Stephenson' country but to survey the route of the railway for atmospheric propulsion. Stephenson was against the atmospheric system and was well versed in speaking on the subject as both the Stephensons, Robert and his father George, viewed the atmospheric system as a revival of the fixed engine controversy, which the elder Stephenson famously dismissed as a 'rope of wind'. Brunel had met Lord Howick in his capacity as chairman of a House of Lords committee, and appears to have been won over by Brunel's enthusiasm for the atmospheric system. The committee had accepted the Croydon Co.'s branch line to Epsom and rejected the rival bid by the South Western on 23 May 1844. It was reported on 31 May 1844 that:

> The great characteristic of this protracted contest is, that the Croydon Company propose to lay down a line from Epsom to London (a length of twenty-one miles) on the atmospheric system. Mr W. Cubitt, Mr Brunel, Mr Samuda, Mr Barry Gibbons, and other eminent engineers, were examined for the atmospheric plan, and Mr R. Stephenson against it.[11]

As the engineer to the Chester & Holyhead Railway (C&HR), Robert Stephenson had been approached by some of the directors to consider its adoption as the motive power on the railway. Stephenson, however, went to considerable lengths to make his own technical investigation into the mode of propulsion that was the atmospheric system. His report outlined seven points supporting his conclusion that the system was unsuitable for long-distance main lines, and the C&HR remained, as Stephenson had intended, a steam locomotive line.[12] The Northumberland Railway offered an alternative route to that proposed by George Hudson and originally surveyed by George Stephenson; the Newcastle & Berwick Railway, a route that would complete the railway between London and Edinburgh. The main objection by Lord Howick to Hudson's route was the impact it would make on his landholdings and that it would spoil the attraction of the family seat, Howick Hall, near Alnwick.[13] This excursion by Brunel into 'Stephenson' territory appears to be the occasion when George Stephenson took him playfully by the collar and told him that he was not to appear north of the Trent![14] It was a good-natured encounter witnessed by Thomas Sopwith, the Newcastle-born engineer then assisting Brunel with the surveys of the Northumberland Railway. Four years earlier Sopwith had taken a somewhat adventurous crossing of the River Avon at Clifton on the iron bar that Brunel had fixed there from one side of the gorge to the other.[15] Whatever the nature of their first professional contact, in September 1844 they were working together: 'Mr Brunel, the eminent engineer, with Mr Sopwith, of this town, are surveying the line of country through which it is proposed to carry the Northumberland (atmospheric) Railway in promoting which Lord Howick has for some time been engaged.'[16]

Notices for the Northumberland Railway's share capital appear from August 1844. Interestingly, the usual railway locomotive and carriage images normally used within advertisements of this kind did not initially figure. However, on 27 September 1844 a locomotive and goods wagon image heads up the notice. The *faux pas* must have been noticed, as in a later advertisement the locomotive block is replaced by one of a goods wagon, probably the closest the typesetters of the *Newcastle Courant* had to depicting the leading, or piston, carriage used on the atmospheric railway.[17] There was local opposition to Brunel's plans; the Newcastle Town Council met to consider the two schemes and the impact on the town and in crossing the River Tyne. For the latter Stephenson proposed a high-level bridge, at an elevation of up to 100ft above the high-water mark. It was considered that one important feature of Stephenson's proposal, Robert having taken over and amended his father's original route, was the opportunity to unite all the railway stations passing through Newcastle in one place, and that the gradients were somewhat easier than those proposed by Brunel, the gradients on Stephenson's line being, of course, designed for locomotive haulage. After discussing the details of the Newcastle & Berwick Railway the following description was given on the plans of Brunel's Northumberland Railway:

> It crosses the river by a bridge ninety-three feet high, passing between the Factory-lane and the plate glass works. It then crosses the Forth Banks (the level of which is to be lowered) at an elevation of 15 feet above the present surface of the ground, it passes nine feet below the present level of Neville-street; under Clayton-street West, and Thornton-street, and from thence by a tunnel 2,410 yards in length, under the road at the Westgate, the Bath-lane, Gallowgate, Leazes-terrace, and the North-road, terminating in Jesmond Fields, from whence the line proceeds direct northward.[18]

The town council gave their assent to Stephenson's line '…as the most desirable for the interests of the town', on condition that the line and the high-level bridge were executed within five years, with the latter accommodating road traffic. Under regulations then recently introduced for railways, the schemes had to go before the Board of Trade who examined both proposals and heard evidence from Brunel and Stephenson. The report of the Board of Trade favoured the Newcastle & Berwick Railway on a number of points, the atmospheric system being, in their opinion, a negative one to that of the established method of propulsion, and that '…its success in a commercial point of view is not yet decided…'[19] Consequently, the committee of the House of Commons decided against Brunel's atmospheric line and threw out the Northumberland Railway's Bill on 26 May 1845. Hudson's route won the day and the atmospheric system had received one of its many setbacks before Brunel himself was to abandon it. Hudson, no doubt, took some satisfaction from this victory, routing the 'outsiders'. Whilst this debate was going on, which appeared to be going their way, Hudson and his supporters seized an opportunity which had come about by chance in stopping the broad gauge from reaching Birmingham. The GWR was negotiating the extention of the broad gauge by amalgamating with the narrow-gauge Birmingham & Gloucester Railway and the broad-gauge Bristol & Gloucester Railway, the Birmingham & Gloucester Railway coming about as a revival, in part, of the proposed

Bristol & Birmingham Railway which Brunel had been involved with in November 1830. In a coup brought about by Hudson's vice chairman on the Midland Railway, John Ellis, a better offer was made and the line from Birmingham to Bristol was snatched from under the noses of GWR.

Despite the disagreements over railway gauge and the battles fought over the atmospheric system when Brunel and Stephenson appeared as witnesses, there was no conflict when the two engineers shared a platform many years later that acknowledged their respective engineering talents. On 24 June 1857 Brunel was awarded an honorary degree from Oxford University, a distinction he had stated he was justly proud of. This was an honorary degree of Doctor of Civil Law which was conferred on Brunel at the annual ceremony known as Encaenia.[20] *The Times* article opened their column on the proceedings that day by reference to the brilliant sun and the cloudless sky ushering in what was the closing day of the Oxford annual festivities. Brunel was one of fourteen to be honoured that year, including Robert Stephenson and the explorer David Livingstone (1813–73). During the degree ceremony, the reception from the assembled crowd to the previous candidate was noted as nothing remarkable but when Brunel and Stephenson were called down there was great excitement and 'loud cheers greeted the two great engineers, Mr. R. Stephenson and Mr. I. K. Brunel'.[21] In terms of the measure of popular applause that the two engineers received on that brilliant and cloudless day in Oxford, theirs was gauged to be the third loudest, with a soldier gaining the greatest acclaim followed by an explorer.[22] The missionary and explorer, Livingstone, who could now be called Dr Livingstone, had become a household name and was hailed as a hero following his return from Africa in 1856. He returned following his discovery, amongst others, of the spectacular falls named in honour of the Queen: the Victoria Falls of the Zambezi.[23] It is fair to assume also that the Revd Baden Powell (1796–1860), the Slavian Professor of Geometry at Oxford, was present at this degree ceremony. A link between Powell and Stephenson can be traced back to Cardiff during the building of the TVR, or, strictly speaking, with Powell's wife and his father-in-law, Captain – later Admiral – William Henry Smyth (1788–1865).[24]

Smyth had been engaged on hydrographic surveys for the Royal Navy, and in 1825 retired to establish a private observatory in Bedford, observing deep sky objects during the course of the 1830s. He came out of retirement during the construction of the first Bute Dock, moving to Cardiff as Lord Bute's advisor and leader of the Bute opposition to the TVR's proposed shipping place on the River Ely[25] – the role of propagandist being added to that of accomplished sailor, celebrated navigator and amateur astronomer.[26] Robert Stephenson had also been engaged by Bute as part of the expert advisors assembled to fight the TVR's proposals and thereby compel them to use Bute's recently opened dock on his terms. Smyth's engagement with Bute led to him meeting Stephenson and the subsequent development of a close friendship that appears to have been shared by Smyth's daughter, Henrietta Grace (1824–1914). Living in Cardiff with her father, she had taken up an interest in painting and, in 1840, depicted the River Taff, a scene that would change ten years later when the river was diverted through a new cut by Brunel.[27] Henrietta married the Revd Baden Powell, and Robert Stephenson was asked to be godfather to their first son who was born on 22 February 1857. This child was christened Robert Stephenson Smyth Powell, but was to become world famous as Baden-Powell, or simply 'BP'.[28] Smyth's 1840 publication under Bute's auspices do him no justice in contrast to his other works, his Bedford observations appearing in 1844 were hailed as one of the most popular treatises for the astronomer, while his dictionary of nautical terms has rarely been out of print.[29] Dr Livingstone was to take Smyth's advice on the best way to enter the 'dark continent' for his next expedition, seeking his opinion 'as to the use of an instrument for determining the Longitude which you recommended'.[30] Smyth's son, Charles

Piazzi Smyth (1819–1900), followed his father as far as astronomy was concerned, and became Astronomer Royal of Scotland.

Stephenson was called upon to assist Piazzi Smyth's scientific work in the years before 'BP' was born, lending him his iron yacht *Titania* so that he could conduct experimental observations. The expedition made his name as a scientist and Piazzi Smyth was duly elected to the Fellowship of the Royal Society in 1857.[31] Robert Stephenson was a keen yachtsman and owned two iron yachts, both called *Titania*, which were designed and built by the naval architect John Scott Russell (1808–82). The first *Titania* was the first iron yacht to use Scott Russell's 'wave-line' form but he had to clip and modify what should have been a true wave-line formed hull in order to comply with the tonnage regulations of the yacht clubs. Scott Russell was the builder of Brunel's *Great Eastern* and as such was implicated, as far as the ship was concerned, in the trials and tribulations of the year 1857 and beyond. There was also a Cardiff connection with Scott Russell as he had also carried out work for Lord Bute.[32] Following his Scottish research work on his wave-line theory Scott Russell had moved to London around 1845 to work on, amongst other things, the editorial desk of *The Railway Chronicle*.[33] The *Titania* was launched, fully rigged, from the Robinson and Scott Russell yard at Millwall on the Thames in early 1850. Stephenson played a role in yachting history with *Titania* by taking up the challenge for the Royal Yacht Squadron (RYS) One Hundred Sovereign Cup in 1851. That year is remembered in yachting history as the year that the yacht *America* won the race round the Isle of Wight, with the cup now becoming famous as 'The America's Cup'.[34] It was Stephenson's second *Titania*, also built by Scott Russell, that was loaned to Piazzi Smyth.[35] Along with the Astronomer Royal, Professor Sir George Biddell Airy (1801–92), Piazzi Smyth would also be consulted by Brunel on the design of gyroscopic equipment for the *Great Eastern*.[36]

Stephenson had already been the recipient of an honorary degree from Durham University in 1850 and in 1857 he was undertaking his term of office as president of the Institution of Civil Engineers (ICE).[37] As the senior vice-president, it was intended that Brunel would succeed him when Stephenson vacated the chair in December 1857 but he was forced to ask for this honour to be held over because of his work schedule and failing health.[38] Joseph Locke, the next vice-president in order of seniority, was proposed and duly elected as the ninth president. Brunel had followed his father in joining the 'civils' and was elected an Associate of ICE on 27 January 1829, the year after it gained its Royal Charter, transferring to Member on 14 June 1831.[39] There were national organisations of professional engineers before this, the oldest being the Society of Civil Engineers, founded in 1771, which became the Smeatonian Society of Engineers, but since its formation in 1818, ICE has grown to become the dominent body and the oldest engineering institution in the world. The name of the former comes from John Smeaton (1724–1792) who coined the term 'civil engineer' in 1761. Henry Robinson Palmer (1795–1844) is regarded as the main instigator of this body and in initiating the Institution at a meeting on 2 January 1818 at Kendal's Coffee House, where he advocated the formation of a society for the advancement of knowledge in civil engineering. In the previous year he had suggested to his fellow engineer, Joshua Field (1786–1863), the idea of founding a society of engineers which would be more accessible to younger engineers rather than the élitist Smeatonian Society of Engineers. In 1820 the leading civil engineer of his day, Thomas Telford (1757–1834), agreed to accept the role of president of ICE. Palmer has strong links with South Wales; his first 'corrugated' iron bridge was erected near Swansea; he built the first dock at Port Talbot in 1836 and had business interests in the Neath area.[40] The London headquarters of ICE, situated at 1 Great George Street, Westminster, has a stained-glass memorial window commemorating Palmer's role which was installed in 1954 and at the time illuminated the main lobby of the building.[41]

Telford, who never joined the Smeatonians, remained president of ICE until his death in 1834. Brunel was an Associate and Member during Telford's presidency, serving on the council in 1834–41, and again in 1845–49, becoming a vice-president from 1850 until his death. In the 1850s a number of regional professional organisations for engineers were being established; the North of England Institute of Mining Engineers had been founded in 1852 in Newcastle-upon-Tyne, followed by the Manchester Association of Employers, Foremen and Draughtsmen in 1856.[42] It was therefore seen as a logical development for South Wales, the moving force behind it being William Menelaus (1818–82), the manager brought in by the resident trustee at the Dowlais ironworks, George Thomas Clark (1809–98). Clark was the former engineer who had worked under Brunel and who would make the acquaintance of Menelaus. In 1855 Brunel wrote to his assistant to say that he would be glad to meet with '…Mr. Menelaus but it can only be in London and after Tuesday next.'[43] From the content of Brunel's reply, Menelaus appears to have been seeking his advice as to what rolling techniques produced the best rail, as far as railway engineers were concerned. Brunel passed on some simple advice: '…that you cannot make a good rail without good materials in the first place.' After going through what he thought were the qualities of a good rail, including those not made of 'the best material', he concluded by offering to send a piece of rail supplied from Dowlais in 1838 and 1839 that was '…some of the best we have ever had and he can examine it'. Menelaus sought to build on the wealth of professional expertise in South Wales and moved to establish such an engineering institute in Merthyr Tydfil, which would be the third such body to be formed in the country.

Following a number of informal meetings, he organised a meeting at the Castle Hotel in Merthyr Tydfil on 29 September 1857 where sufficient support was forthcoming to launch a scheme for an engineering institute. The South Wales Institute of Engineers (SWIE) would retain its original title for 150 years, although the title had initially been proposed but not accepted as the 'South Wales Institute of Mining and Mechanical Engineers'at the first meeting held at the Assembly Rooms of the Castle Hotel.[44] On 29 October 1857, following the proposals developed by the provisional committee, the South Wales Institute of Engineers was founded, and Menelaus was elected as first president. At its first council meeting in 1858, a proposal was debated for inviting certain eminent engineers to become honorary members of the institute. Robert Stephenson and Brunel were put forward in this respect but it was decided to postpone the matter until the institute had matured and shown that it merited recognition.[45] This was not the only connection with Brunel as many of the members had associations with the engineer during his work in South Wales; Thomas Forster Brown, the president in 1873, was a partner with Samuel Dobson and joint engineers with John Hawkshaw (1811–91) of the 1863 Penarth Dock (where Henry Marc Brunel would be employed early in his career as an assistant engineer). Henry Keyes Jordan (SWIE president in 1883) was born in Clifton, Bristol, in 1838 and became articled to Scott Russell of Millwall where he worked on the construction of the *Great Eastern*. Scott Russell's son, Norman, would would also join SWIE when he came to Cardiff to build iron ships at the Bute shipyard in 1864 (see Chapter 8). Another president, in 1863–64, was Alexander Bassett, a civil and mining engineer based in Cardiff, who claimed to have had '… considerable experience with the late Mr. Brunel in conducting tunnel operations.'[46]

John Lean (1818–78) was another of Brunel's assistants working on the SWR and VNR who was also a SWIE member. Thomas Joseph (1819–90), a colliery prorietor, was vice-president of SWIE in 1882. He began his career managing the Plymouth Ironworks of Anthony Hill at Merthyr and, when giving evidence on a parliamentary Bill, was proud to say that he was '… for several years a member of Brunel's staff.'[47] He had worked on the VNR on the

Right: *William Menelaus (1818–83), the Dowlais Ironworks manager who promoted the South Wales Institute of Engineers, becoming first president in 1857. (Courtesy of Merthyr Tydfil Reference Library)*

Far right: *Sir John Hawkshaw in 1877. (SKJ collection)*

Merthyr to Hirwaun section and when he died was one of the oldest members of SWIE.[48] One engineer who held the office of president on no less than two occasions, both of two successive years, was William Thomas Lewis, later Baron Merthyr of Senghenydd (1837–1914). An article on Lewis in 1884 had claimed that when an apprentice at the Plymouth ironworks he had undertaken sketches for the rolling of Brunel's rails, and that he was brought into contact with Brunel.[49] However, a couple of days later a letter by a T. Creswick of Swansea refutes several aspects of the article, not least that he was an apprentice at Plymouth (as Hill never took apprentices) and 'In reference to Brunel is another error. All drawings went through my hands, if not actually made by myself.'[50] His father, Thomas William Lewis, also a member of SWIE, was the engineer at Plymouth (or millwright, to quote Creswick) and that T.W. Lewis was articled to his father, in which capacity he no doubt met Brunel with his father, T.W. Lewis having the advantage of meeting Brunel '… during that great engineer's visits to Mr. Hill, when the first rails for the Great Western Railway Co. were rolled at the Plymouth Ironworks.'[51]

A fellow engineer on good terms with the 'triumvirate' was George Parker Bidder (1806–78). Bidder, known as the 'Calculating Boy' due to his extraordinary mathematical abilities, began his professional career in Cardiff working for the Ordnance Survey.[52] He would go on to work for Stephenson, becoming a much respected and talented engineer, and was to succeed Locke as president of the ICE. Bidder had met Brunel as an adversary in the 'battle of the gauges' and '…found him a very dangerous and most able adversary', but at the same time Bidder stressed that he was equally high-minded, and above the use of mean and pettifogging practices to accomplish his objects. 'He was one of those men in whose minds private friendship was never interrupted by professional strife.'[53] The extent of such friendships are not evident when it came to working with his assistants. We can examine the letters written by Brunel but in many cases the other side of the correspondence is missing, as in the situation with George Bush, his assistant on the TVR, and although we know that William George Owen (1810–85) asked Brunel to act as his son's godfather, there is little personal detail.[54] Owen was a competent engineer; after Brunel's death he succeeded him as engineer of the South Wales Railway and remained on as engineer for the South Wales section of the GWR with the merger of the SWR, before becoming chief engineer of the GWR in 1868. Similarly, it is difficult to gauge the relationship with Robert Pearson Brereton, Brunel's senior assistant who would complete much of his unfinished work after 1859. There are exceptions when dealing with men like George Thomas Clark and Daniel Gooch who emerge out of Brunel's shadow, and his relationships with architects seems to have been on a more equal footing and he was prepared to work alongside them, despite the 'sibling rivalries' of the professions. This was despite the fact that the interface of the railway with the public (i.e. the public architecture of stations and other works) was very much from a previous age when compared to the technology of the railway.

Reasons for this were understandable; the promoters of the early railway had wished to overcome any adverse perceptions by the public as to the safety and reliability of this new mode of transport and to present an impression of solidity that only 'classical' forms of architecture could project. Architects commissioned by the railways drew upon a number of stylistic revivals of the classical forms of architecture, from the Gothic, the Italian Renaissance to the Byzantine-Romanesque. There was safety in using tried and tested forms of architecture, particularly with the Victorians' interest in the cultures of ancient civilisations and medievalism. Hence the style chosen for the grand portico and Great Hall built by the London & Birmingham Railway at Euston, a point of entry to the railway described by Charles Dickens, where Major Dombey began his railway journey. The scale of the Doric portico, or propylaeum, presented the equivalent to the Victorian city of '… what the city gate was to the ancient city'.[55] The fluted columns that supported the entablature and pediment at Euston were higher than any other contemporary London building, and, whilst it drew nothing from the state of the art that the railway represented, it was a powerful symbol of railway achievement. Such use of past civilisation's architecture was not necessarily a reaction against industrialisation but an attempt by the promoters of the railway, as already stated, to impress the public and also to assure their shareholders. Some architects such as Augustus Welby Northmore Pugin (1812–52) criticised the uselessness of such ornaments, particularly when contrasted with the simple train sheds beyond the monumental edifices.[56] The London & Birmingham Railway, with Robert Stephenson as its engineer, employed the architect Philip Hardwick (1792–1870) to provide its monumental grand entrance at Euston along with the Ionic portico erected at the other end of the line at Curzon Street, Birmingham. Architecture was an area that Brunel naturally took a great interest in, an important part of his engineering work, for example in his first independent commission; the Clifton Suspension Bridge.

Taking shape within the same timescale as these on the GWR were the principal stations on the TVR at Merthyr and Cardiff which included train sheds which were most likely to have been designed by George Bush under Brunel's direction. They were not to the same standard as Temple Meads; indeed, the original wooden buildings of Cardiff Station were described as little more than that of a glorified fowl house whilst the intermediate stations on the line were basic affairs.[57] Apart from bridges and other structural works, South Wales could claim few architectural works outside his station buildings, most of which were constructed in timber. A more substantial building on the TVR was the three-storey Bute Dock Station building now known as Cardiff Bay Station. In stucco with a hipped, overhanging roof, it is the oldest in the city but there is uncertainty over its actual date of construction and designer. The original section was put up a few years after the opening throughout of the TVR in 1841 and therefore may or may not be by Brunel, and it was extended by a two-storey extension at the southern end in the 1860s.[58] Materials and building technology were not advancing at the same rate as the processes of the railway, and as shown by the GWR and TVR examples above, station architecture either belonged to a previous age or was of basic timber construction. The railway had yet to develop a modern architectural style for the age as there was no 'railway' or even 'Victorian' style for this element of the railway; that is, until the beginning of the 1850s with the emergence of a completely new architectural style heralded in with the building of the Crystal Palace. A milestone in the development of modem architecture, it was one that was most associated with typical Victorian architecture although not all of the details of the Crystal Palace were Victorian and its development can be clearly traced back to previous work such as the architectural conservatory, the development of which was stimulated by the availability of good quality sheet glass from 1833 and the removal of a prohibative tax on the same in 1845.[59]

One man, Joseph Paxton (1803–65), building upon the designs of architects like Decimus Burton, arrived at an innovative and revolutionary design that was not only suitable for the Great Exhibition of 1851 but one which presented a new architecture for the railway.[60] Matthew Digby Wyatt (1820–77) served on the executive committee of the Royal Commission for the Great Exhibition, and included Robert Stephenson, who would also serve on the building committee. Other members on this were Brunel and the architect Charles, later Sir Charles, Barry (1795–1860), and it was chaired by William, later Sir William, Cubitt (1785–1861). The building committee received 245 designs for the proposed building but ended up turning them all down and coming up with a design of their own, Brunel's contribution to which was a great cast-iron dome on a huge brick-built structure. The logistics of constructing such a building in time and the cost ruled it out, and the whole project appeared in jeopardy, and then Paxton sketched his revolutionary glass and iron structure. This would later be dubbed the 'Crystal Palace' and it made such an impression that everyone including Brunel supported its adoption to house the exhibition. The connection with Wyatt led to Brunel's most famous partnership with an architect. Brunel was to play an important role in bringing Paxton's design to acceptance for the Great Exhibition and would use this new 'architecture' himself. It is unnecessary to go into detail here regarding Paddington. Suffice to say it was one of the most influential station designs of the Victorian age and the first to be built of cast iron, wrought iron and glass to the Paxton or unit construction method.[61] Brunel asked Wyatt to join him as his assistant:

Right: *Sir Matthew Digby Wyatt in a portrait painted in 1870. (Courtesy of the Royal Institute of British Architects)*

Below: *The through station and new entrance façade at Bristol Temple Meads opened in 1878 and was one of the last architectural works of Matthew, later Sir Matthew, Digby Wyatt (1820–77). The curved train shed to the right is a wrought-iron roof structure by Francis Fox (1818–1914). (SKJ collection)*

Now in this building which *entre nous* will be one of the largest in its class I want to carry out strictly and fully all those correct notions of the use of metal which I believe you and I share (except that I should carry them still further than you) and I think it will be a nice opportunity. Are you willing to enter upon the work *professionally* in the subordinate capacity (I put it in the least attractive form at first) of my *assistant* for the ornamental details?

Wyatt was born on 28 July 1820 at Rowde, near Devizes, Wiltshire. The family background boasted a number of architects and artists and included his eldest brother Thomas Henry Wyatt (1807–80). Wyatt was to marry the second daughter of Iltyd Nicholl of The Ham near Llantwit Major in Glamorgan on 11 January 1853, Mary (*d.*1894). He began his career in his brother's office and made a name for himself in industrial design. In 1849 he was employed by the Society of Arts to accompany Henry Cole to Paris and report upon the French exhibition of that year. His report helped to foster interest for a British exhibition on the same or larger scale and he was selected for the post of secretary to the executive committee of the Great Exhibition. He superintended the execution of Paxton's building in Hyde Park and would present a paper on the construction of the exhibition building to the ICE, for which he won a Telford Medal.[62] His career was a prolific one and his circle of friends included Edward Lear. Amongst his domestic work he would rebuild The Ham, near Llantwit Major, for his in-laws, the Nicholls. Towards the end of his life he was living nearby at Dimlands Castle (a house that also belonged to the Nicholls), where he had retired to recover his health, but he died there on 21 May 1877. Through the Nicholl family connections he was buried at Usk churchyard, his brother Thomas Henry Wyatt designing a tomb of pink granite. After his collaboration with Brunel, he would be involved in other railway commissions, notably the Patna railway bridge over the Sona in India from 1855 (with J.M. Rendel). The new through station at Temple Meads, in collaboration with Francis Fox (1818–1914), received his attention in 1871. Fox had been one of Brunel's SWR resident engineers and later became the engineer for the Bristol & Exeter Railway, in which capacity he was responsible for the design of the single-span curved roof of the B&ER's through train station at Temple Meads, completed in 1877.[63] Wyatt would have been living at Dimlands when one of his former pupils, the Gothic architect and designer William Burges (1827–81), started his monumental work for the 3rd Marquis of Bute at Cardiff Castle and Castle Coch.

The son of a Swansea-based omnibus proprietor was to make his mark for a later generation of railway travellers with the distinctive GWR 'art deco' buildings of the 1920s and '30s. This was Percy Emerson Culverhouse and his work includes the former Arrivals Side Offices at Paddington, a steel-framed and clad building designed by him in 1935.[64] As well as engineering and architectural links there were the designers and manufacturers, men like William George Armstrong, later 1st Baron Armstrong of Cragside (1810–1900), who was to supply hydraulic equipment for Brunel's railways and docks. In this cateogory there is the one instance where Brunel, the engineer who frequently stood alone, asked George William Lenox to 'stand with him'; but on this occasion he declined Brunel's invitation. The Brown Lenox chainworks at Newbridge, Pontypridd, had been founded by Captain Sir Samuel Brown RN (1774–1851), the man who had turned down Brunel's earlier request to tender for the Clifton chain links. In 1857, with Lenox now in charge, they agreed to undertake work for Brunel on the *Great Eastern*, and the engineer was to visit the works at Pontypridd several times during the course of work on this order, the largest size of 2⅞in easily being the largest chains made up to that date. The chains were made by Newbridge and, after testing, were transported down the Glamorganshire Canal by barge and then by coastal vessel around to the Millwall chainworks of Brown Lenox, next door to Scott Russell's yard. Lenox was invited to the Millwall Shipyard

The Ham, West Side, Llantwit Major.

The Ham, Llantwit Major. The original house had been the main seat of the Nicholl family since the six-teenth century, but by the 1850s it was felt to be somewhat modest for their situation. With Wyatt marrying Mary Nicholl in 1853, the family took advantage of having an architect amongst them and work on a new Victorian Gothic mansion began in 1861. It was gutted by a fire after the Second World War and demolished in the 1960s, although some remains of the once magnificent gardens, such as the loggia and the ornamental pond, can still be seen. (SKJ collection)

Situated near Llantwit Major, Dimlands was built by the Revd Robert Nicholl sometime after 1800. Wyatt retired to Dimlands towards the end of his life, and died there in 1877. He is buried at Usk where the Nicholl family also had consid-erable interests, the house being demol-ished in the 1950s. (SKJ collection)

during the first launch attempt and Brunel had arranged for the chain to be wound around two large wooden drums and used to control the descent of the *Great Eastern* into the water during the sideways launch. The position of the drums, however, offered a convenient back-drop to the photographs taken by Robert Howlett. Brunel asked Lenox to stand with him in one of the photographs to be taken by Howlett with the chains as a background. The reluc-tance of Lenox to come into a photograph meant that his image would, as far as we know, never be seen on the ground glass screen of Howlett's plate camera, and one of the resulting images is regarded as one of the most famous in photographic history.[65]

CHAPTER 2 NOTES

1 Believed to have been said by Brunel to George William Lenox, see Jones, Stephen K. (1980), p.43, *A Link with the Past: the History of the Newbridge Works of Brown Lenox and Co., Pontypridd*, in Denning, Roy (Ed.), *Glamorgan Historian, Vol.12* (Stewart Williams: Barry). J.P.M. Pannell also writes in his book, *Man the Builder* (1977), that Brunel wrote this on the back of the Brown Lenox copy of the Howlett photograph (p.113). However, the author examined this particular photograph and could find no inscription on the back; it is believed that another copy, once in the possession of Brown Lennox, may have been inscribed.

2 This has been covered in Vol.II. The proposed railway was the Worcester & Porthdinllean Railway.

3 See Vol.I.

4 Stephenson, however, was unable to support Brunel at Chepstow but made up for it at the *Great Eastern* launch.

5 Watson, Garth (1988), p.42, *The Civils, The Story of the Institution of Civil Engineers* (Thomas Telford: London).

6 Bailey, Michael R., Ed. (2003), p.162 and pp.229–230, *Robert Stephenson – The Eminent Engineer* (Ashgate Publishing: Aldershot). *The Times*, 26 June 1857, recorded the continuing debate on the mapping of Scotland.

7 University of Bristol, *Special Collections, Brunel Letters*, Book 5, pp.334–335, I.K.B. to Henry Robertson, 11 January 1848.

8 Rolt, L.T.C. (1959), pp.73–5, *Isambard Kingdom Brunel* (Longmans, Green & Co.: London). Captain William Moorsom engineered the revived scheme in 1836, and his route was adopted, which included the Lickey Incline.

9 This was the new or revived railway direct from London to York, which the *Newcastle Courant* of 12 April 1844 believed was the work of Brunel.

10 This was Henry George Grey (1802–94), who became the 3rd Earl Grey on the death of his father in 1845.

11 The *Newcastle Courant*, 31 May 1844, reporting from *London News* of 29 May on the committee of the House of Commons with Lord Howick in the chair.

12 Bailey, Michael R., Ed. (2003), pp.198–199.

13 Bailey, Michael R., Ed. (2003), pp.109–110.

14 Sopwith, Robert (1994) p.216, *Thomas Sopwith, Surveyor – An Exercise in Self-Help* (The Pentland Press: Bishop Auckland).

15 Sopwith, Robert (1994) p.174. See also Vol.I, p.86.

16 The *Newcastle Courant*, 13 September 1844.

17 Advertisements appearing on 27 September and 11 October 1844.

18 The *Morning Chronicle*, 17 January 1845.

19 Reported in the *Leeds Mercury*, 1 March 1845.

20 *The Times*, 19 September 1859, p.7. This was in an extended obituary notice on the engineer, which *The Times* referred to as a distinction he was justly proud of. I am grateful to Simon Bailey, Keeper of the Archives at the University of Oxford, for additional information.

21 *The Times*, 25 June 1857.

22 The greatest applause was given to Sir William Fenwick Williams (1800–83), 'the hero of Kars' during the Crimean War, followed by David Livingstone. *The Times*, 25 June 1857.

23 It was while Livingstone was recuperating from his quest to find the source of the Nile that he was 'found' in 1871 by the Welsh-born explorer; Henry Morton Stanley (1841–1904).

24 See Vol.I and II.

25 Smyth was to publish his opinion on the TVR proposals in his *Nautical Observations on the Port and*

Maritime Vicinity of Cardiff in the autumn of 1840. See Vol. I, pp. 171–2.

26 It was while he was in Cardiff that Smyth prepared the observations that he had made at his former observatory in Bedford, England, for publication. The observations of 850 objects were published in 1844 as *A Cycle of Celestial Objects*, but became more widely known as the Bedford Catalogue. Further biographical details about Smyth are given on the website of SEDS (Students for the Development of Space) and in the Royal Astronomical Society's 1866 obituary at the NASA ADS Service.

27 Cardiff Central Reference Library has two versions of the watercolour by Henrietta Grace Smyth (later Mrs H.G. Baden-Powell), see Vol. II, colour plate 19, 'Cardiff: a view of the River Taff as it flowed in 1840…'.

28 Born at 6 Stanhope Street (now 11 Stanhope Terrace), Paddington, London, BP was the sixth son and the eighth of ten children of the Revd Baden Powell. His mother added Baden, his father's first name, to form a hyphenated surname in 1869 in order to improve his social standing.

29 Smyth, W.H., *The Sailor's Word-Book: The Classic Dictionary of Nautical Terms* (Conway Maritime Press).

30 David Livingstone to Admiral Smythe [sic], 18 January 1858, letter in the Wellcome Library, London, with a copy in the National Library of Scotland, MS. 10779 (15).

31 Provided with a £500 grant from the Admiralty and private donations of instruments from the 'scientific men of Great Britain', Piazzi Smyth travelled to Tenerife in Stephenson's 140-ton yacht, *Titania*.

32 In 1864 Scott Russell returned to Cardiff to build an iron ship, but as he was now bankrupt, the enterprise was managed by his son Norman Scott Russell. See Chapter 8.

33 Emmerson, George S. (1977), pp. 29–31, *John Scott Russell: A Great Victorian Engineer and Naval Architect* (John Murray: London). See Chapter 1 regarding his railway editor role.

34 There was much interest in the match that took place on 29 August 1851, a race between schooners of wood and of iron. On the return passage, *America* flew away from *Titania*, winning by a huge margin of forty-seven minutes and forty-eight seconds, a decisive victory. *America* was sold two days later.

35 The first *Titania* caught fire and burnt out in May 1852 but, being built of iron, her frame survived the fire and Scott Russell rebuilt her as the *Themis*, while at the same time a new *Titania* was built for Robert Stephenson which was launched in 1853.

36 Buchanan, R.A. (2002), p. 120, *Brunel, The Life and Times of Isambard Kingdom Brunel* (Hambledon & London: London).

37 Bailey, Michael R. Ed. (2003), p. 237. Brunel was elected a Fellow of the Royal Society in 1830 (after he became a member of the Institution of Civil Engineers), some nineteen years before Stephenson's election as an FRS.

38 Buchanan, R.A. (2002), p. 219.

39 The actual transfer form is missing; it is not with the Associate form or filed in sequence in 1831. His obituary notice (Obituary ICE Minutes of Proceedings, Vol. 19, 1859–60, pp. 169–173) says transfer was in 1837 but it is recorded in the Ordinary Meetings Minutes in 1831. Information supplied by Carol Morgan, Archivist of ICE.

40 The bridge consisted of three arches, two of them of 48ft span and one of 50ft span. Other work included the improvement of the dynamometer and the invention of the suspension railway, and from 1826 he was appointed resident engineer to the London Docks.

41 Ironically, considering that the work of Palmer tends to be overshadowed today, much of the light behind this window has since been blocked out by developments behind the ICE headquarters. Watson, Garth (1988), p. 226.

42 The North of England Institute of Mining Engineers added 'and Mechanical' to its title in 1866, and gained its Royal Charter in 1876, becoming the Federated Institution of Mining Engineers in 1889. The Manchester Association of Employers, Foremen and Draughtsmen became the Manchester Association of Engineers in 1885.

43 Letter dated 30 November 1855, from Brunel to Williams, J., [sic] in Elsas, Madeleine (1960), p.178, *Iron in the Making, Dowlais Iron Company Letters 1782–1860*, Glamorgan Quarter Sessions and County Council and Guest Keen Iron and Steel Co. Ltd, Cardiff. Clark's connection has been covered in the two previous volumes.

44 In 1878 it moved to Cardiff, and gained its Royal Charter in 1881. In 2007 SWIE became an educational trust; the South Wales Institute of Engineers Educational Trust (SWIEET). The author had the privilege of being elected to SWIE in 2007 – the last new member.

45 The South Wales Institute of Engineers, Sesquicentenary Brochure 1857–2007 (SWIE: Cardiff).

46 *Western Mail*, 19 October 1869.

47 *Western Mail*, 28 March 1882. Joseph was giving evidence on the Glyncorrwg, Rhondda & Swansea Junction Railway Bill to the Select Committee of the House of Lords.

48 *Western Mail*, 11 July 1890.

49 *Western Mail*, 2 August 1884. An article entitled 'Captains of Industry' by Major Evans Rowland Jones. It was claimed by the Merthyr historian Charles Wilkins (1831–1913) that Brunel and Hill were related.

50 *Western Mail*, 6 August 1884.

51 *Western Mail*, 9 February 1900, reporting the death of T.W. Lewis of Abercanaid House, Merthyr Tydfil.

52 Watson, Garth (1988), pp.192–3.

53 Watson, Garth (1988), p.43.

54 See Vols.I and II for further information on these relationships. I did Owen an injustice in Volume II by referring to the recommendation to his employment being made by Brunel's clerk Bennett; the recommendation came from the civil engineer and contractor George Hennet (1799–1857).

55 Betjeman, John (1972), p.124, *London's Historic Railway Stations* (John Murray: London). Quoting from Meeks, C.L.V. (1957), *The Railway Station: an Architectural History* (Architectural Press: London).

56 Pugin was, like Brunel, the son of an émigré French architect who came to England to escape the Revolution.

57 *The Railway Times*, p.138, quoting from Francis Whishaw's unpublished work. The intermediate stations of Llandaff, Pentyrch, Taff's Well, Newbridge and Navigation House (Abercynon) were described as being '… executed in an economical style.' See Vol.I, p.187.

58 Biddle, Gordon (2003), p.577, *Britain's Historic Railway Buildings* (Oxford University Press: Oxford).

59 There were also structural issues that had to be overcome in an iron and glass construction, namely the difference in expansion of the two materials which cause the putty holding the glass to crack and fall away with consequent corrosion. Hepper, F., Nigel (Ed.) (1982), p.62, *Kew: Gardens for Science and Pleasure* (HMSO: London).

60 It was a structure of great utility and yet embodied philosophical and ideological implications on the use of glass as an architectural material.

61 Vaughan, Adrian (2006), p.140, *Brunel, An Engineering Biography* (Ian Allan Publishing Ltd: Hersham). Adrian Vaughan claims that it was the first station in the world to combine industrial functionalism with grandeur.

62 Minutes Proceedings of the Institution of Civil Engineers, 1851, Vol.10, pp.127–165.

63 See Vol.I, p.178, for an incident in 1849 concerning Brunel and Francis Fox. The through station was further extended on the east side in the 1930s by P.E. Culverhouse.

64 The father was no doubt involved with contracting work for the SWR and other railways. I am grateful for the family history on P.E. Culverhouse researched and supplied by Anne Gardner.

65 No photographic likeness of Lenox is known to exist; only a bas relief bust has been attributed to him.

3

SEVERN GATEWAY

'I SHALL HAVE THE SATISFACTION OF BRIDGING THE SEVERN, AS WELL AS THE TAMAR'[1]

On the night of 25 October 1960, a group of workmen gathered around a wireless set in the signal box at Severn Bridge Station. The set was tuned to the BBC light programme in order to pick up the world bantamweight championship being broadcast that night.[2] The workmen, taking a break during a major contract of bridge refurbishment, listened intently to the commentary of the fight between Freddie Gilroy and Alphonse Halimi that had begun at 9:15 p.m.[3] The fight went to fifteen rounds and at about twenty-five minutes past ten a crash was heard − not that of a boxer hitting the canvas but the devastating sound of bridge pier 17 being knocked down and taking with it two spans of the bridge. The debris of the cast-iron twin-cylinder pier and two wrought-iron 171ft spans crashed down on top of two self-propelled fuel barges, the MV *Arkendale H* and the MV *Wastdale H*, igniting petrol vapour that sent a flame 250ft in the air.[4] Five of the eight crewmen onboard the two vessels perished in the conflagration that engulfed their barges and spread across the water, the River Severn being ablaze for almost a mile upstream.[5] The two barges should have entered Sharpness Docks to proceed up the canal to Gloucester but in the descending fog both barges missed the navigation lights for the dock entrance and with the approaching high tide the river was flowing strongly, pushing the barges along at a rate of around 4 knots. In an attempt to get back to the dock entrance but hampered by the dense fog, both vessels collided and locked together, striking the pier minutes later with a combined force of 858 tons.

The death toll could have been far higher if the workmen had not decided to take their break when they did, and consequently there were no casualties on the bridge itself. It had been hit many times in the past but the cost of any substantial protection around the bridge piers had always been considered prohibitive. The bridge was never repaired although it would not be finally dismantled and removed until 1969, and the remains of the barges still lie in the Severn off Purton as a reminder of that fateful night.[6] Completed in 1879, the Severn railway Bridge crossed the river above Sharpness and carrying a single line, and was perhaps not as badly situated as Brunel's earlier proposal roughly five miles below this. With the rejection of Brunel's bridge it was the first crossing of the Severn below Gloucester to be realised. The promoters of this bridge were able to overcome objections to the bridge interfering with the clear navigation of the river although the very risk which had been levelled at Brunel's bridge would be the cause of its destruction; a 'bridge strike' by a barge. The tragic end of the Severn Railway Bridge was a story of foul weather, bad luck and bravery, but above all a consequence of 'impediment to navigation', the implications of which should have been taken more seriously.

The River Severn has presented both a barrier and a gateway to South Wales since time immemorial. Before the crossings of the Victorian era the estuary of the Severn divided England and Wales as effectively as the straits of Dover separated France from England.

Cardiff is separated by only ten miles of water from Weston-super-Mare but over land the distance is nearer 100 miles. As the Afon Hafren, to give it its Welsh name, the Severn flows from the slopes of Plynlimon Fawr for some 210 miles through Wales and England, at places forming the border between the two, before discharging into the Severn Estuary and the Bristol Channel. On its way through the hills of central Wales, the Severn takes a meandering northerly course, spanned by many historic bridges, until reaching Shrewsbury where it turns south. Before this town, with its famous 'Welsh' and 'English' bridges, the Severn flows under the Holyhead road at Montford and the first bridge designed by Thomas Telford.[7] Continuing southward, the Severn passes under more historical bridges including the famous Iron Bridge of 1779 at Coalbrookdale, and then under four of Telford's cast-iron and masonry bridges.[8] The last of Telford's bridges can be found downstream of Gloucester at Over which, when opened in 1830, was the lowest permanent crossing of the Severn. Despite all the bridges, smaller river craft were able to navigate up the Severn to Shrewsbury whilst Bewdley represented the upper reaches for larger vessels. The Severn would be opened up even further with the construction of canals such as the Staffordshire & Worcestershire Canal which connected with the Severn at Stourport and gave access to the Black Country via the Trent & Mersey Canal.

As a strategic waterway and the greatest in medieval Britain, the Severn carried a greater volume of traffic than any other river in Europe before the development of canals. It had a unique range of river obstructions, shifting sands, dangerous tides and currents, but was not prone to silting up due to the great tidal range of the Bristol Channel.[9] The Severn was a 'free river' in that anyone could travel or convey his goods along its whole length without being subject to toll. Coasting trade had been carried on for centuries between the ports and shipping places of the Welsh coast and the 'Welsh metropolis', as Bristol was often called. The

Opened in 1966, the Severn Bridge was to make redundant the car ferry that operated on the Old Passage route. Remains of the now derelict ferry buildings at the Beachley landing stage can be seen in this 1979 photograph. (SKJ photograph)

*The Severn Railway Bridge was second only to the Tay Bridge in length when it opened in 1879, having a total length of 1,387 yards (including approaches). The work of constructing the twenty-one river spans had to be undertaken using timber staging. (*The Severn Tunnel, Its Construction and Difficulties 1872-1887, *by Thomas A. Walker, Third Ed., 1891)*

golden age of this network was in the period before the railways when all possible use was made of coastal transport with manufacturers going to great lengths to transport their goods in this way. For example, in 1775 the Horsehay Co. of Wellington, near Shropshire, were sending pig-iron to Chester, a journey that began with transport by cart to the Severn.[10] It would be loaded onto Bristol-bound riverboats and there transhipped to vessels sailing around the coast of Wales and up the Dee to arrive at Chester. Transport by this route involved a journey of over 400 miles by sea and two trans-shipments, compared to sixty miles by land, and well-illustrates the state of the roads and the reluctance of carriers to use them. There would be road improvements, of course, before the construction of the railways, but they had little impact on the transport of heavy goods and merchandise.

Ferry traffic across the Severn, principally at the Old and New Passage crossings, would also be affected as such ferries were not for the faint-hearted and passengers were more than happy to transfer to the alternative offered by the railway. Before these alternatives the ferry crossings were important and well-established routes, although exactly when they first came into prominence is unrecorded. Roman legions are said to have used the Old Passage route on their way between Caerleon and Silchester, and this may be where the name Aust comes from – a derivation of Augustus after Trajectus Augusti or the ferry of Augustus.[11] Before the railway option Defoe had described the Old Passage crossing, from Aust across to Beachley at the mouth of the River Wye, as being an '…ugly, dangerous and very inconvenient ferry over the Severn…'[12] Travelling to Wales in 1725, Daniel Defoe decided the alternative route via Gloucester was the safest and surest way, taking account of the weather and seeing the sorry state of the ferry boats at Aust:

> …the sea was so broad, the fame of the Bore of the tide so formidable, the wind also made the water so rough, and which was worse, the boats to carry over both man and horse appeared so very mean, that in short none of us cared to venture: so we came back, and resolved to keep on the road to Gloucester.

It would take 100 years from Defoe's observations before any significant improvements would be made to the Old Passage or Aust crossing (incidentally the two main crossings are incorrectly named as the New Passage was the older of the two). This passage was closed following the marooning and subsequent drowning of parliamentary troops on the English Stones in 1645.[13] From Beachley, on the opposite side of the Severn, the route through South Wales followed the prehistoric track known as the 'portway', which in its eighteenth-century road form was in a poor state of repair. Calls for improvements were made repeatedly and a local association was formed to press for action but it was not until 1823 that the Postmaster General appointed an engineer of the calibre of Thomas Telford to advise on improvements to the mail coach routes.[14] This was part of the planned improvement of mail services to the south of Ireland with the Post Office taking control of the mail packet service and replacing the sailing packets with steamers between Milford Haven and Waterford in 1824.[15] Obviously the Severn was seen as the major obstacle on the route to Milford Haven and Telford examined the crossing at New Passage and the longer crossing from Uphill Bay in Somerset to Sully Island, although an additional twenty-two miles of new road would have to be constructed for the latter. The Post Office would not entertain the cost of the latter proposal and the former involved the crossing of the English Stones section of the river, which in Telford's opinion was:

> One of the most forbidding places at which an important ferry was ever established, a succession of violent cataracts formed in a rocky channel exposed to the rapid rush of a tide which has scarcely an equal on any other coast.[16]

As Telford had himself rejected New Passage he now considered the Old Passage, which was after all the narrowest point on the estuary. While he was carrying out this work his two suspension bridges were in the course of construction in for the Holyhead Road in North Wales, so it came as no surprise that he would suggest a similar solution in 1824 for the Severn at the Old Passage. A suspension bridge here would be more or less on the line of the first Severn Bridge which would be opened 142 years later, but the suggestion was not taken forward. The South Wales route to Milford Haven and Southern Ireland did not have the political support that the Holyhead route to Dublin enjoyed, and the idea of a road bridge was revived several times during the nineteenth and earlier twentieth centuries.[17] Later that year, on 26 November 1825, the same newspaper announced that 'Mr Telford recommends the Old Passage'. Such a recommendation had been long expected as it was really the only option left. Earlier in 1825 *The Times* reported on the formation of the Old Passage Ferry Association at a meeting held on 7 April at the Beachley Inn.[18] It marked the opening of a new era with new facilities to be constructed on both sides of the Severn, the Beachley slipway to be constructed a short distance south of the old landing place below the passage house. A steamboat, intended to tow two large boats, was to be purchased by the association which would cross every fifteen minutes, taking between ten and twenty minutes according to the state of the wind and tides. Henry Habberley Price (1794–1839) was to be responsible for the construction of platforms and stages for the convenient shipment of horses and carriages.

There had also been talk at the meetings of buying up the rights to the rival New Passage ferry to ensure sufficient custom which, not unnaturally, led to an outcry from the New Passage Co. who responded by ordering a steamboat. Their steam packet, the *Saint Pierre*, was the first steamship built at Newport (by Pride and Williams) and would be launched and running before the end of 1825, almost two years before the steamboat of the Old Passage Association entered service. Their first steamship was the *Worcester*, built by William Stott of

Bristol, however, by 1831 the improved facilities, which included the landing slipways at the Old Passage, began to tell, and mail coaches stopped using the New Passage ferry. In that year the *Saint Pierre* stopped running and the following year Old Passage gained a second steam-ship, the *Beaufort*, built by Richard Watkins of Chepstow, although a number of sailing boats continued to be used on the crossing.[19] Despite these improvements the crossings were still treacherous in bad weather, so much so that the Post Office considered the installation of a chain ferry in 1836.[20] James Meadows Rendel (1799–1856) surveyed both the Old and New Passages but reported that tidal currents were too strong for his system, which employed a steam engine on the ferryboat to pull, or warp, itself along a submerged chain. Steam engines in wooden ships were also a constant danger and *Worcester* was burnt to the waterline follow-ing an accident at the end of 1837, being replaced by a second ship with the same name the following year.[21] Steam, however, was safer than sail when it came to navigating the Severn, a fact highlighted in the tragic disaster that occurred on 1 September 1839 when a sailing barge capsized and all onboard were drowned, including William Crawshay III, the son and heir of the ironworks empire.

Despite these and other incidents, crossing the Severn was seen as an accepted risk of trav-elling by men such as David Lewis of Stradey Castle who would use such means until the railway provided an alternative. Lewis was a supporter of Brunel's railways and rode on the footplate with him when the SWR was opened through Llanelli, on its way from Swansea to Carmarthen, in October 1852. But to get to Bristol in May 1844 he chose to drive to Newport and take the steamer across from there, leaving at 9.00 that morning but not getting to Bristol until 4.15 that afternoon, which he described as a 'long passage'.[22] Other Severn ferries included the ancient ferry from Newnham to Arlingham where it had been proposed to establish a bridge and a chain ferry before the Second World War.[24] With the completion of the SWR allowing through communications to Haverfordwest in 1854, the railway was now opening up the coastal ports and towns bordering the Severn and the Bristol Channel. As a result sea traffic lost its competitive edge and Bristol its monopoly.[23] The construction of the SWR had physically curtailed another ancient ferry at Purton Passage by trapping Gatcombe Quay on the wrong side of the railway embankment.[24] By 1855 the use of steamboats on the Old Passage could not be justified and another disaster that year threatened to close the ser-vice altogether.[25] However, the ferry continued and it was advertised in 1858 that 'ordinary' or sailing boats were being used, although the steamboat could be ordered in advance by letter. Both *Beaufort* and *Worcester (2)* were scrapped at the end of the 1850s with the loss of traffic as a consequence of the railway's advance. At the Old Passage there was no incentive to persevere with steam sailings, and eventually any sailings at all, and the passage was closed, to be revived for the motorcar in the 1920s.[26]

With the completion of the Gloucester & Deanforest Railway from Gloucester to join the SWR and the Bristol & Gloucester railway, both sides of the Severn could now be accessed by rail; but not content with this, railway interests now wished to advance directly across the Severn. There were attempts even before this, dating back to tramroad days when various individual proprietors of the Bullo Pill Railway Co. promoted the construction of a tunnel under the Severn in 1810. They acquired the rights of the Newham Ferry and began to con-struct a road tunnel that would also allow the access of adapted tramroad wagons. Like Marc Isambard Brunel's Thames Tunnel some fifteen years later, a major influx of the river brought work to a close, causing a long-drawn-out but temporary setback for Brunel's undertaking but an unrecoverable setback for what was the first Severn Tunnel.[47] Other proposals included the ambitious twenty-arch railway viaduct of 1834 by Charles Blacker Vignoles (1793–1875), and both he and Brunel considered the possibilities of tunnelling under the Severn. Between the

Steam tug Ceres *at the entrance to the Glamorganshire Canal. In the background can be seen the* Hamadryad, *a former warship being used as a sailors hospital. A landing stage here was used by the Severn ferry boats carrying passengers such as Brunel. (Courtesy of Cardiff Central Library)*

Admiralty rejection of the SWR timber viaducts and later proposals such as Sir John Fowler's viaduct of 1865 (rejected as an 'impediment to navigation'), one particular mode of crossing was to be promoted which would not be seen as such an impediment. Such a crossing started to take shape in Brunel's mind after 1844 when he considered bringing together the convenience of the railway and the flexibility of the steam ferry: a line running from Bristol to the banks of the Severn, across which the traffic, passengers and light goods would be carried by a steamer to a short branch from the SWR on the other side. The first prospectus for such a railway was the Bristol & South Wales Junction Railway (B&SWJR) and one of which Brunel was not initially the engineer, but nevertheless it was a scheme that attempted to address the setback of not bridging the Severn.[28] The engineering position was fulfilled by George and John Rennie, but Brunel would replace them by the time the Monmouth Extension Prospectus was published in June 1845.

The B&SWJR proposed to construct a railway, with ferries, from Bristol to join the SWR, and it obtained its Act on 26 June 1846.[29] This Act allowed for the construction of a railway, with ferries, from Bristol to join the SWR and to buy up the rights to, initially, the New Passage ferry, with a further Act in 1847 to buy up the Old Passage ferry. There were difficulties over legal titles to property and ferry rights and in raising share capital and arguments led to the abandonment of the company in October 1853. It appears that the B&SWJR operated the ferry at New Passage for a period from 1847 until sometime before the company was wound up, but did so at a loss. Brunel was consulted by Henry Somerset, the 7th Duke of Beaufort (1792–1853), on the future of the ferries, the family now owning the rights to the Old Passage ferry. His reply to the duke's agent on 28 January 1853 expressed the view that neither the GWR or SWR companies would be likely to buy the Passage Ferry but that it was (in his personal opinion) in the interests of both companies to get it worked as soon as possible.[30] No movement on this appears to have been made for the following year or so later and Brunel was consulted again, this time by the new Duke of Beaufort.[31] Work was well under way on the River Tamar Bridge at Saltash, and Beaufort asked if such a bridge could be built at the Old Passage. Brunel replied, again to his agent, on 30 May 1854:

I should be very glad if the Duke thinks seriously that it would benefit his interests, to look seriously into the question and give the best advice I can. And if I should be able to suggest a feasible plan and there should be friendly people ready to make it, I shall have the satisfaction of bridging the Severn, as well as the Tamar.[32]

It was probably from this that Brunel would keep the possibility of 'bridging the Severn' in mind as in April 1857 when sketching a Severn bridge with obvious similarities to Saltash, and to which he would pose the question 'Severn Bridge. Q: is 1,100ft practicable?'[33] In his letter to Beaufort's agent Brunel believed that within fifty years there would be a bridge or tunnel across the estuary – he would be right on both counts but sadly he would not have the satisfaction of bridging the Severn. A year after the demise of the B&SWJR another proposal called the Bristol South Wales & Southampton Union Railway (BSW&SUR) was the subject of meetings and the engineer Thomas Evans Blackwell was appointed at the first meeting of the Provisional Committee on 3 November 1854. One of the provisional directors and honorary secretary was Adam Jack, the Bristol agent for Coalbrookdale Co., a company that had long used the strategic waterway of the Severn to carry their iron. Plans were deposited in November 1854 for an ambitious scheme that included steam ferries, or 'floating steam bridges' as they were called, in which railway wagons and coaches would be transported on the ferry boat. Such ferries were already in use in Scotland; indeed, the first train ferry in the world, dating from 1849, used the specially designed ferry boat *Leviathan* across the Firth of Forth between Granton and Burntisland. It was designed by Thomas Grainger in 1849 and built by Robert Napier & Sons, the loading piers and apparatus designed by Thomas Bouch (1822–80). Blackwell went up to study the operation of this system in which up to twenty goods wagons could be loaded straight on the *Leviathan*.[34] Passenger traffic was taken on separate paddle steamers but for the Severn it was intended to also take passenger coaches across on the ferry, with passengers having the choice of staying onboard or moving to a separate passenger deck. The scheme failed to secure a sufficient take up of shares for a Bill to be presented in the session of 1854/55 and another attempt was proposed. By this time there was strong Welsh support for the scheme, with Christopher Rice Mansel Talbot (1803–90), the chairman of the SWR, and Thomas William Booker of the Melingriffith works onboard.[35]

Before the Bill could be introduced, however, the House of Commons needed to be satisfied that the interests of river users were not compromised, and a public enquiry was ordered by the Admiralty which opened on 25 March 1857 at the New Passage Hotel. Opposition was raised by the City of Gloucester and the Gloucester & Berkeley Canal Co. as well as owners of small craft and the steam packet *Wye* which ran between Chepstow and Bristol. There was also a suspicion that a chain ferry would be used for the ferry even though Rendel had dismissed the Severn as being unsuitable back in 1836. Brunel was questioned at some length here but stated that he never intended that the ferry would be operated by the use of submerged chains and the confusion may have been made with the term 'floating steam bridges' to describe the intended ferry. Finally the Admiralty would give assent to the scheme in principle and a revised proposal was suggested by Brunel who replaced Blackwell as engineer in 1855. Southampton was to be dropped from the title, and as the Bristol & South Wales Union Railway (B&SWUR) an Act was obtained on 27 July 1857, but the B&SWUR would not be as ambitious as the Firth of Forth crossings in being a train ferry, and it would therefore be confined to passenger and light goods only.[36] It would not be completed until after Brunel's death. In *The Life of Isambard Kingdom Brunel*, a crossing that was eventually built as the B&SWUR is referred to as one that had been for a long time contemplated, and that:

…Mr. Brunel devoted much time to a careful consideration of the Severn in order to determine the most suitable point for the crossing. He decided that the best place would be at what is known as the New Passage. The arrangements had to be made in accordance with the requirements of the Admiralty. Trains run to the end of timber piers extending into deep water, and there are staircases and lifts leading to pontoons, alongside which a steamer can come at all times of tide. The tide at this part of the Severn rises 46 feet.[37]

The steam ferry would utilise steam boats and several engineers would be associated with the construction of the B&SWUR: in addition to Robert Pearson Brereton who would complete the work after Brunel's death, there was the resident engineer Charles Richardson (1814–96) and the contractor Rowland Brotherhood (1812–83). Both were well-known to Brunel and apart from engineering connections there was another common factor between the two engineers as both enjoyed the game of cricket. Indeed, Brotherhood had enough children of his own to form a cricket side and would play other family teams. The railway development of Victorian Britain greatly boosted the popularity of games like cricket where previously, because of the difficulty of travelling, few fixtures were played away from the home pitch. The landed gentry and industrialists encouraged the creation of cricket clubs, the latter as a way of encouraging their workmen to pursue healthier activities and stay out of the alehouses. Brunel was indirectly responsible for the siting of Cardiff Cricket Club on the reclaimed land and existing river meadow at the rear of the Cardiff Arms Inn.[38] The land had been reclaimed when Brunel diverted the River Taff for the SWR with the Marquess of Bute, as landowner, allowing the club to use it at a peppercorn rate. Lord Bute was to become the patron of the club at Cardiff Arms Park and in 1876 the park would be shared by a rugby team who would eventually take over the site now dominated architecturally by the Millennium Stadium. Association football, like rugby, was to become a great spectator sport with the opportunity for the fans, and the team, to travel by railway to away games, and the first professional team, Sheffield F.C., was founded in 1857. Cricket clubs such as Cardiff and Newport on the Welsh side of the Severn Estuary could play teams on the English side by using the frequent sailings of the steam packet but sometimes there would be an unfair advantage as in 1844 when many of the English visitors suffered from seasickness during a rough crossing![39] The *Great Britain* also had the distinction of taking the first cricket tourists, the famous 'Eleven of All England' team, to Australia in 1861.[40]

Richardson's passion for the game of cricket would bring him into conflict with Brunel, who rebuked him for his lack of energy and activity in attending to railway work, a complaint levelled at him because of his '…alleged devotion to amusement and amongst other things to cricket…'[41] If the option had been open to him Richardson might have forsaken engineering for cricket, but his talent for engineering was fostered by his mother, his father having died when Charles was six years old. Born and brought up in Cheshire, he was educated privately in England and France, going on to attend Edinburgh University up to the year 1833 where he demonstrated skills and ability in mechanics. He appears to have been taken on as a pupil of Sir Marc Isambard Brunel according to 'an esteemed Swansea correspondent' who stated that 'He was a pupil of Sir Isambert [sic] Brunel, and was employed by him in making the Thames Tunnel.'[42] This experience no doubt was important enough for him to be taken on by Brunel and employed on the Box Tunnel and on other sections of the GWR. His work here may have brought him into contact with Brotherhood, working under his father, William Brotherhood (1778–1839), then excavating the foundations of the Wharncliffe viaduct at Hanwell.[43] With a lull in work he was sent to check rails being produced in Ebbw Vale and Merthyr, frequently staying at the Cardiff Arms.[44] After this he worked on the Sapperton Tunnel of the Cheltenham & Great Western Railway (C&GWR) and it was on the C&GWR

Dated, Jan. 11, 1864. · 3061

TAFF VALE RAILWAY COMPANY.

THE Directors of this Company are prepared to issue DEBENTURE BONDS for 5, 7, or 10 Years, under the powers of "The Taff Vale Railway Act, 1857." Applications to be made to the undersigned.

By Order of the Board of Directors,

EDWARD KENWAY,

Cardiff, 20 January, 1864.   Secretary.
                                              3094

BRISTOL AND SOUTH WALES UNION RAILWAY.

On and after 1st JANUARY, 1864, THROUGH BOOKING between BRISTOL and all STATIONS on the SOUTH WALES LINE, *via* the BRISTOL and SOUTH WALES UNION LINE, which is the nearest by Fifty-six miles each way.

THIRD CLASS BOOKED THROUGH by all Third-class Trains on the SOUTH WALES LINE, Up and Down.

REDUCED FARES:—

	ORDINARY.			RETURN.	
	1st	2nd	3rd	1st	2nd
	Class.	Class.	Class.	Class.	Class
	s. d.	s. d.	s. d.	s. d.	s. d.
NEWPORT or CARDIFF to BRISTOL, or *vice versâ*	5 0	3 6	2 0	7 6	5 0
CHEPSTOW, ditto, ditto..	3 6	2 6	1 6	5 0	4 0
SWANSEA, ditto, ditto..	13 1	9 7	5 9½	21 6	15 6

Other Stations in proportion.

For particulars see Time Bills, which may be obtained at the Railway Stations, and at the Office of this Paper.
3040

Above: *Charles Richardson (1814–96). Richardson talked about the B&SWUR as being intended to convey coal across the river in trucks, but that Brunel decided to open firstly for passengers. (*The Severn Tunnel, Its Construction and Difficulties 1872–1887, *by Thomas A. Walker, Third Ed., 1891).*

Left: *An advertisement in* The Times *(22 January 1864) listing fares on the recently opened B&SWUR.*

that Richardson incurred Brunel's criticism for his love of cricket. However, Richardson must have mended his ways because five years later Brunel invited him to be the resident engineer on the B&SWUR, selling the prospects to him:

> I want a man acquainted with tunnelling and who will with a moderate amount of inspecting assistance look after the Tunnel with his *own eyes*, for I am beginning to be sick of Inspectors who see nothing, and resident engineers who reside at home … The country immediately north of Bristol and Clifton etc. I can't vouch for any cricketing but I think it highly probable.[45]

Brotherhood had engaged in some speculative work by surveying at his own expense a line from Bristol Temple Meads through Queen Square to cross the Floating Harbour by an iron bridge.[46] The line then went across Canons Marsh and through a tunnel under St Vincent's Rock at Clifton Down to run alongside the Avon to its mouth where it proceeded behind the sea wall to New Passage. This is the first BSW&SUR proposal, but one that was not supported, and Leleux refers to its supporters suddenly opposing it with Brotherhood then having to resurvey the route, this time running north from Bristol. Following the passing of the B&SWUR Act in 1857, Brotherhood, who was well known to Brunel, was awarded the contract in September 1858, and work began on the curiously named, considering the distance from that place, Almondsbury Tunnel, with a heavy cutting at Horfield being started a month later. This tunnel was 1,245 yards long and is now known as the Patchway Old or Down Line Tunnel. The single line railway was eleven-and-a-half miles long from the junction half a mile east of Temple Meads, and included stations at Lawrence Hill, Stapleton Road, Filton, Patchway, Pilning and New Passage.[47] The gradients being as steep as 1 in 75 for two miles up to Filton and 1 in 68 for one-and-a-quarter miles down to Pilning, on the Welsh side a branch line less than a mile long (seventy-four chains) ran from Portskewett to the pier. The piers were the most innovative part of the undertaking as they incorporated floating pontoons at the end of each of the timber piers, to which the trains would run with access by stairs and lifts to the pontoons. The piers extended far enough out to be in sufficient water for the steam ferry boat at any state of the tide with the pontoons floating with the tide and therefore at the same level as the boat when it came alongside. The New Passage Pier was 1,635ft long and Portskewett Pier 708ft long, and hotels on either side catered for the needs of passengers. The Bristol line opened on 8 September 1863 and the Portskewett branch on 1 January 1864, but there was no opening ceremony until 25 August 1864 and the first crossings did not take place until 7 September, and then not to the public until the following day.

The steam packet boats were provided by the B&SWUR and operated under contract by John Bland, the GWR working the Bristol line and also the Welsh branch as the SWR had been absorbed on 1 August 1863. In 1868 the B&SWUR was absorbed into the GWR and the Portskewett branch was converted to standard gauge in May 1872 with the remainder narrowed in August the following year.[48] Brotherhood also invested in the B&SWUR, taking shares to the value of £2,500, with his son Peter Brotherhood (1838–1902) taking £250. Brotherhood would later have to take more shares in part payment amounting to a quarter in shares and debentures for work which cost over £240,000, and eventually he held 1,740 £25 shares. In March 1869 the bank foreclosed on his assets, which included his Chippenham Railway Works, and he was forced to assign to his creditors and the bank his B&SWUR shares.[49] Unfortunately for Brotherhood, these shares would soar in value within a short time of the enforced sale due to the future implications of the B&SWUR line in the GWR's Severn Tunnel plans. Following this, Brotherhood left Chippenham and took up a post as general manager for Herbert and Charles Maudslay at the Bute Ironworks, the works used by Norman Scott Russell to build the first large iron ship in Cardiff. Brotherhood moved with his wife and three of his sons to Cardiff, living at 9 Charles Street. Edward Finch of Chepstow was also to be involved here according to a newspaper advertisement taken out on

The southern portal of the Patchway down line, the original B&SWUR line to New Passage, which was referred to as Almondsbury Tunnel. A plaque to the chairman and Brunel was erected here and, until it was moved to Temple Meads in 1986, was probably one of the most inaccessible plaques ever erected on a railway line! See colour section, picture 9. (SKJ photograph)

Rowland Brotherhood (1812–83) and his family cricket team, photographed in about 1860. (Courtesy of Sydney Leleux)

1 September 1869 in which he states he had taken the Bute Ironworks, Cardiff and '… will make as heretofore, Iron Bridges, Caissons, and Dock Gates of Iron …' [50] This would account for the large numbers of dock gates built, with 'watertight compartments', as Finch had built the semi-buoyant wrought-iron dock gate for Brunel's Briton Ferry Dock in 1861. The yard supplied a total of some sixteen wrought-iron dock gates for the docks at Newport, Cardiff and Bristol, and this and other work lasted until 1874 when the Maudslay brothers decided to close the yard. Brotherhood moved back to Bristol and he was successful in tendering for Severn Tunnel work, taking contracts for the sinking of two shafts and driving headings east and west on the Monmouthshire side. [51]

In 1858 Richardson was appointed resident engineer under Brunel on the B&SWUR and following Brunel's death in 1859 the responsibility of his role increased. It was while working on the ferry piers in 1862–63 that he began to pursue the idea of a Severn Railway Tunnel. It took ten years to get such a proposal accepted as the plan was initially rejected on the grounds of cost, one attempt for an Act was withdrawn in 1865 in preference to John Fowler's South Wales & Great Western Direct Railway, a double mixed-gauge line of forty-one miles from Wootton Bassett to Chepstow. [52] This was to cross the Severn at Oldbury Sands by a viaduct up to two-and-a-quarter miles long and 100ft above high water, but it would never be built. Daniel Gooch, then recently returned to the GWR as chairman, saw it as an extravagance that really benefitted the moving force behind the project, one of the GWR directors, Richard Potter: '…he had an object beyond the Co's interest in the matter: his coal company thought the line would be a benefit to them.' [53] Gooch was to write that he had soon put an end to this scheme as far as the GWR were concerned and without their support the scheme was abandoned and the company dissolved in 1870. Richardson addressed numerous meetings and carried out much lobbying, even approaching contractors of the calibre of George Wyles and Thomas Brassey. The latter

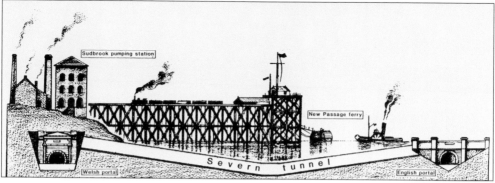

Top: *Portskewett Pier showing the Severn Tunnel works at Sudbrook in the background.* (The Severn Tunnel, Its Construction and Difficulties 1872–1887, *by Thomas A. Walker, Third Ed., 1891)*

Middle: *Drawing showing the Portskewett Pier with the floating pontoon and ferry boat and a representation of the tunnel beneath the Severn. (Graphics by Owen Eardley)*

Left: *This photograph was taken on the site of the Portskewett Pier in 1979. Some timber stumps and a GWR boundary (now still gone) can be seen. (SKJ photograph)*

OPENING OF THE BRISTOL AND SOUTH WALES UNION RAILWAY AND FERRY : NEW PASSAGE HOTEL AND PIER.

Opening of the New Passage Pier and facilities of the B&SWR as reported in The London Illustrated
News. *(SKJ collection)*

indicated that he would build a Severn Tunnel; however, he recognised the risks and added that
at least £500,000 should be available for contingencies.[54] Richardson was not the only promoter
of a tunnel scheme as in 1869 three such proposals went before the public. His was described as a
tunnel under the river at New Passage to connect the '...Bristol & South Wales Union line with
the South Wales section of the Great Western, and thus avoid the present ferry.'[55]

The cost of Richardson's tunnel was estimated at £730,000, a cost based on the fact that no
new railways would be required excepting for about a mile on each side to allow for a gradual
fall to the level of the tunnel. Other schemes put forward were by Mr Fulton for a tunnel near
Chepstow (and a railway to Wootton Bassett) and one about a mile and a half above Lydney.
This would join with the Midland Railway near Berkeley road and the GWR near Stonehouse,
and was proposed by James Abernethy and Alexander Bassett, a Cardiff-based engineer who had
worked with Brunel. All of these schemes were driven by the desire of the colliery proprietors to
find an alternative route to transport South Wales steam coal to English markets. In 1869 South
Wales coal was being conveyed to London on the narrow gauge via Worcester and Didcot. Bassett's
mixed-gauge route would be some thirty-eight miles shorter but it failed to gain sufficient support. Fowler was at this time consulting engineer for the GWR and was to bring forward another
bridge scheme in 1871, which also saw a number of competing bridge concepts. Eventually,

Right: *This photograph was taken of the site of the Bristol & South Wales Union Railway Hotel in 1979. (SKJ photograph)*

Below: *The B&SWUR ferryboat, PS* Christopher Thomas. *(Grahame Farr collection)*

however, Richardson's scheme was taken forward, and in 1872 the Severn Tunnel Bill was passed by Parliament and construction began in the following year. Richardson was to be chief engineer with John, later Sir John, Hawkshaw (1811–91) as consulting engineer, and his principal assistant was to be Gooch's third son Alfred (1846–87), who, like his father, became an engineer and his father found him work on the *Great Eastern* during her first, unsuccessful, cable-laying outing. In 1872, no doubt again because of his father's influence, he was working in the GWR engineers' office at Oxford before taking up the post and moving to live at Ashfield House in Chepstow.

The full story behind what was to be the longest railway tunnel under water in the world, which would turn into a classic example of the engineer's fight against adversity, cannot be told here apart from a brief mention of those engineers linked with what was surely the leviathan of all Victorian railway tunnels. Hawkshaw was a highly accomplished engineer with many engineering achievements and, as we have already seen, was involved or connected to Brunel or his works since 1839 or even earlier. A tunnel under the English Channel was one of his projects which, although never completed, involved the young Henry Marc Brunel.[56] In October 1879 the headings from the English and Welsh sides were only 138 yards apart when a great inrush of water occurred. With the work brought to a standstill the GWR brought in Hawkshaw as engineer-in-chief and Richardson's role was much diminished. Hawkshaw insisted on bringing in Thomas Andrew Walker as contractor.

*The Severn Tunnel portal on the Welsh side (*Brunel and After: The Romance of the Great Western Railway, *published by the GWR in 1925)*

Walkers earlier tender for the construction of the tunnel had been rejected by Richardson who believed his price of £1 million was too high and could be done for 25 per cent less by direct labour.[57] Walker had worked under Brassey and undertook contracts at home and abroad. Domestic works included the Metropolitan and Metropolitan District Railways, the Manchester Ship Canal and Barry Docks. He came to the attention of Hawkshaw when working on the extension of the East London Railway, from the end of Marc Isambard Brunel's Thames Tunnel at Wapping, to Shadwell. This railway, running in a tunnel which continued underneath the London Docks which Walker had to construct by building two coffer dams in the dock itself, was commented on by Hawkshaw as a display of Walker's great skill in combating difficulties.[58] Walker brought the best men from these works to the Severn Tunnel and Hawkshaw appointed A.G. Luke to represent him at the works as his resident engineer.

Gooch had lost confidence in Richardson; this was evident following a visit to the tunnel in May 1878, when he confided to his diary that the tunnel was a very anxious job and that 'Richardson, the engineer, has no go in him and does not move without consulting me, making me almost the engineer.' Gooch was present at the opening of the Severn Railway Bridge on 17 October 1879, the day when, unbeknown to him at the time, the Great Spring broke through. He invited those at the gathering to walk through the completed tunnel, thought to be possible in only a few weeks time. Little did he realise the truth of the joke he made: '…It will be rather wet and you had better bring your umbrellas!'[59]

The Severn Railway Bridge was a more conventional form of crossing, designed by George William Keeling and George Wells Owen, Owen being the son of William George Owen. Thomas E. Harrison was to act as consultant, and the bridge was opened four years later. The Severn Railway Bridge was 4162ft long, a single line viaduct consisting of twenty-one spans supported on huge cast-iron cylinders and situated half a mile above the entrance to Sharpness Docks and the canal to Gloucester. Keeling and Owen had made a survey of the river as early as 1859 and in 1865 submitted a proposal for a bridge crossing at Purton allowing both road

The contractor of the Severn Tunnel, Thomas Andrew Walker (1828–89). As a contractor, his Welsh works also included the Prince of Wales Dock at Swansea, the Penarth Dock extension, Barry Dock and Railways, including Wenvoe Tunnel and the Dinas Powys to Cogan junction of the Cardiff, Penarth & Barry Junction Railway. Walker was to settle a few miles from Sudbrook at Mount Ballan and was buried at Caerwent churchyard where there is a Lych-gate to his memory. (SKJ collection)

and rail traffic on broad- and standard-gauge metals.[60] This did not proceed through lack of financial support but another scheme for a crossing at Sharpness was put forward in 1871 which proceeded to Parliament and for which an Act of Parliament was obtained on 18 July 1872, work not beginning until 1875 with a foundation stone being laid on 3 July of that year.

Back at the tunnel it was not until the end of 1881, with the most strenuous efforts of the engineers, contractors and divers, that the tunnel was pumped dry and work was allowed to continue. On 5 September 1885 a special train with Gooch, Hawkshaw, Walker, Frederick Saunders of the GWR and staff from the works such as Alfred Gooch and other special guests including ladies and Lady Gooch, made the first train journey through the tunnel.[61] As some works had yet to be completed it was expected that the tunnel would be open in two or three months but it was not until the end of 1886 that the tunnel was opened to passenger and goods traffic, even though a special coal train had run through from Aberdare to Southampton on 9 January 1886. Delays in the opening were caused by Hawkshaw's insistence that sufficient pumping facilities should be in place. The construction of the four-mile, 628-yard-long tunnel took almost fourteen years at a cost of more than £2 million. It is to Richardson that credit is due for the conception of the scheme and he developed a system for successfully aligning the two headings with unparalleled accuracy. Even Gooch acknowledged that some of the decisions made by Hawkshaw, going against Richardson's advice, such as the deepening of the tunnel and the adoption of linings of halfbrick rings instead of the vertical bond of brickwork specified by Richardson were perhaps not the right ones.[62] That was not the extent of his ingenuity; returning to his love of cricket, Richardson was the inventor of the cane-spliced cricket bat and a catapult for bowling. He was elected a member of the Institution of Civil Engineers in 1875 and in his later years wrote three papers on the Severn Tunnel, one on arches and one on the disposal of sewage. At a half-yearly meeting of the GWR in February

When steam pumping was decommissioned, Swansea Museum (Royal Institution of South Wales) took possession of one of the 20-ton, 30ft-long, wrought-iron beams in August 1961. Like the Cornish beam engines there, it was built by Harvey & Co. of Hayle in 1879. (SKJ photograph)

1889 the sum of £1,000 was voted to Richardson in recognition of his exceptional services in the execution of the tunnel.[63] He died at his home, 10 Berkeley Square, Bristol, on 10 February 1896, survived by his wife, Mary Frances, and two sons. When the costs were finally reckoned the tunnel came in at two and a half times the original estimate and nearly double Walker's contract price, at £1,806,248.

Walker stated in the preface to his book that 'One sub-aqueous tunnel is quite enough for a lifetime'[64] – a sentiment no doubt shared by all those engaged in such undertakings, such as Richardson and Marc Isambard Brunel. Less than a year after the tunnel opened Alfred Gooch was dead and Walker was to pass away on 25 November 1889 at his home in Mount Ballan, situated a few miles from Sudbrook. He died of Bright's disease at the age of sixty-one – the same illness that took Brunel at the age of fifty-three.[65] The steam ferry was to ply back and fore across the Severn without major incident, although a serious fire engulfed the Portskewett Pier on the night of 22–23 May 1881. It was discovered several hours after the last train had departed and there was suspicion that it was the act of men on strike at the railway works, but this was refuted by Walker.[66] Workers on the other side of the Severn at New Passage saw the blaze and the steam ferry boat *Christopher Thomas* crossed back over and attempted to put out the fire using its pumps, but despite this and the efforts of the Newport fire brigade and a Bristol crew and tender brought over the next morning, much of the pier was destroyed. A reduced ferry service continued to operate and an alternative service was run over the Severn Railway Bridge before the full service was resumed on 16 June 1881. Seven crossings a day and two on Sundays were timetabled towards the end of the B&SWUR's operation. The opening of the Severn Tunnel on 1 December 1886 for passenger traffic marked the end of Brunel's unique steam ferry railway, with the last crossing by the steam packet *Christopher Thomas* taking place the evening before.

CHAPTER 3 NOTES

1 Norris, John (1985), p.6, *The Bristol & South Wales Union Railway* (Railway and Canal Historical Society: Oakham).

2 My account was inspired by a story on the Severn Railway Bridge disaster in the Gloucestershire features webpage: http://www.bbc.co.uk/gloucestershire/focus/2004/10/sharpness_rail_disaster. shtml. It stated that the workmen had chosen to take a break as Henry Cooper was fighting in a heavyweight boxing match; however, Henry Cooper was not fighting that night.

3 The fight was won on points by the Frenchman Alphonse Halimi.

4 An investigation into the cause of damage to the Severn Railway Bridge, pp.69–78, by Peter Mason in *The Structural Engineer*, Vol.41, No.2, February 1963.

5 The *Arkendale H* of Bristol was carrying 296 tons of fuel oil and the *Wastdale H* of Hull was carrying 350 tons of petrol. See article by Jamie and Susan Davies; 'A Tale of Two Tankers', in the October 1992 issue of *Waterways World*, pp.58–61.

6 See Vol.II, pp.78–82.

7 Montford consists of three masonry arches, two of 50ft (15.3m) and a central arch of 58ft (17.7m).

8 Iron Bridge was designed by Thomas Farrols Pritchard for William Derby III. There were originally five Telford bridges which included the first of his cast-iron bridges at Buildwas (replaced in 1905).

9 This, the second highest rise and fall in the world, scours the river's mouth, preventing a build-up of silt.

10 Simmons, Jack (1962), p.32, *Transport: A Visual History of Modern Britain* (Vista Books: London).

11 Jordan, Christopher (1977), pp.14–15, *Severn Enterprise: The Story of the Old and New Passage Ferries* (Christopher Jordan: Ollveston).

12 Rogers, Pat (Ed.), p.128 (1992).

13 Cromwell ordered the closure of the ferry, and it reopened in 1718 as the 'New' Passage.

14 The South Wales Association for the Improvement of Roads called for road improvements in 1789. *The Cambrian* for 17 May 1823 reported that Telford was to survey the roads from Carmarthen to Bristol and Gloucester.

15 Farr, Grahame (1967), pp.40–42, *West Country Passenger Services* (T. Stephenson & Sons Ltd: Prescot).

16 Jordan, Christopher (1977), p.22.

17 Finally the Severn Road Bridge was built between 1961 and 1966 to carry the M4 motorway. A steel-cabled suspension bridge with a central span of 3,240ft, it was designed under the direction of Sir Gilbert Roberts by Freeman, Fox & Partners and Mott, Hay & Anderson. Telford had another attempt to promote his suspension bridge design as part of a new road from Briton Ferry to Swansea and Llanelli and, as reported in *The Cambrian* on 25 June 1825, involved a large span crossing over the River Neath.

18 *The Times*, 22 April 1825.

19 Farr, Grahame (1967), pp.99–100. *Saint Pierre* was sold to Joseph Tregelles Price (brother of H.H. Price) and converted to a sailing vessel in 1841. The Old Passage Ferry Association had given way to a new arrangement by the time the Beaufort was acquired.

20 The first 'floating bridge' was designed by Rendel for the River Dart in Devon and opened in 1831. There are some links between the Brunel and Rendel families, see Chapter 10.

21 The new *Worcester* was an iron paddle steamer supplied by James and William Napier of Glasgow.

22 *Diary of David Lewis*, 31 May 1844, courtesy of Sir David Mansel Lewis of Stradey Castle, Llanelli.

23 Not all of which was under GWR control because the Bristol & Gloucester Railway had passed into Midland Railway ownership in 1846.

24 See Vol.II, p.84. Willis, Margaret (1993), pp.70–72, *The Ferry between Newnham and Arlingham* (Alan Sutton: Stroud). Keeling, and later his engineering practice, Keeling & Reichenbach, submitted a

number of bridge proposals here. A permanent crossing was attempted with wartime technology in 1948. This was with a floating bridge made up of hexagonal steel 'lilly pads', but it was not a success and the crossing gradually went out of use.

25 Another sailing vessel, the *Despatch*, heavily laden with sheep, pigs, oxen, a horse and fourteen passengers, capsized in April 1855. Seven drowned on this occasion with stories that one survivor came ashore at Severn Beach by hanging on to the tail of one the beasts!

26 Enoch Williams, the main instigator behind the Newham proposals, reopened Old Passage in 1926 and operated car ferries there from 1931 until the opening of the Severn Bridge in 1966.

27 See Vol.I re: Thames Tunnel and Vol.II re: Severn Tunnel of 1810.

28 *The Times*, 9 April 1845.

29 See Vol.II, p.72. Its share capital was set at £250,000 with power to borrow £83,333. No subscription by the SWR was authorised to this line.

30 Norris, John (1985), p.6, *The Bristol & South Wales Union Railway* (Railway and Canal Historical Society: Oakham).

31 Somerset, Henry (1824–99). Before the age of fifteen he held the title of Lord Glamorgan and then Marquess of Worcester before succeeding to his father's title.

32 Norris, John (1985), p.6.

33 University of Bristol Special Collections, large sketchbook 11, f.42, 20 April 1857. See Vol.II, p.83.

34 Marshall, John (1971), p.54 and 189, *The Guiness Book of Rail Facts and Feats* (Guiness Superlatives Ltd: Enfield). A similar service operated across the Firth of Tay from 1858.

35 His uncle Richard Blakemore had to be 'accommodated' by Brunel in order to overcome his objection to the 1836 Taff Vale Railway Bill. As Booker-Blakemore (on the death of his uncle in 1855), he supported the SWR and was chairman of the Cardiff Steam Navigation Co. See Vol.I, p.106.

36 Vol.II, p.72.

37 Brunel, Isambard (1870, reprinted 1971), pp.89–90, *The Life of Isambard Kingdom Brunel* (Longmans, Green & Co.: London, 1870, reprinted by David & Charles, Newton Abbot, 1971).

38 See Vol.II, pp.113–116. See also Parry-Jones, David, Ed. (1984), pp.37–38, *Taff's Acre, A History and Celebration of Cardiff Arms Park* (Willow Books: London).

39 Hignell, Andrew (2008), *Cricket in Wales*, p.19 (University of Wales Press: Cardiff).

40 Fogg, Nicholas (2002), pp.148–150, *The Voyages of the Great Britain* (Chatham Publishing: Rochester).

41 Buchanan, R.A. (2002), p.159, *Brunel, The Life and Times of Isambard Kingdom Brunel* (Hambledon & London: London).

42 This article in the *Western Mail* for 9 September 1885 is an attempt by one of his oldest friends to put the record straight on Richardson's role on the Severn Tunnel.

43 Leleux, Sydney A. (1965), p.10, *Brotherhoods, Engineers* (David & Charles: Dawlish). The main contract for the brickwork of the viaduct was let to Grissel & Peto. See Chapter 3 for other references to Brotherhoods work as a contractor.

44 Vol.II, p.116.

45 Buchanan, R.A. (2002), p.159.

46 Leleux, Sydney A. (1965), p.15.

47 MacDermot, E.T., revised by Clinker, C.R., pp.2–3 (1964). Ashley Hill station was added in 1864.

48 Barrie, D.S.M., The Bristol & South Wales Union Railway, in the *Railway Magazine*, pp.423–427, Vol.79, December 1936.

49 'Being thrown on the market created a glut and only fetched £4 to £5 each.' Leleux, Sydney A. (1965), pp.26–29.

50 *Western Mail*, 1 September 1869. Finch, as a member of the late firm of Finch & Heath, was forced to make such a move as the relative of his partner, the late Mr Heath, had decided to realise their interests.

51 Leleux, Sydney A. (1965), pp.47–48.

52 MacDermot, E.T., revised by Clinker, C.R., p.15 (1964), *History of the Great Western Railway, Vol.II* (Ian Allen: London).

53 Wilson, Roger Burdett, Ed. (1972), pp.111–2, *Sir Daniel Gooch Memoirs and Diary* (David & Charles: Newton Abbot).

54 Cowles, Roger (1989), p.12, *The Making of the Severn Railway* (Alan Sutton: Gloucester).

55 *The Times*, 25 October 1869. Both the B&SWUR and SWR were part of the GWR by this time.

56 See Vol.I and II for the Brunel connections with Hawkshaw.

57 *The Times*, 27 August 1884. This would bring the cost closer to his original estimate of £730,000.

58 Walker, Thomas Andrew (1891, reprinted 1969 by Kingsmead), pp.14–15. The 'Introductory Note' by Hawkshaw was written for the third edition, after the death of Walker.

59 Platt, Alan (1987), p.185, *The Life and Times of Daniel Gooch* (Alan Sutton: Gloucester).

60 Huxley, R.M. (1984), p.10, *The Rise and Fall of the Severn Bridge Railway* (Alan Sutton and Gloucestershire County Library: Gloucester).

61 *The Times*, 7 September 1885.

62 When building the B&SWUR, Richardson discovered a source of brick clay at Cattybrook, near the west end of Patchway Tunnel, and the brickworks he set up here supplied almost 20 million bricks to the tunnel.

63 *The Times*, 20 February 1889.

64 Walker, Thomas Andrew (1891, reprinted 1969 by Kingsmead), p.10.

65 This disease is now called nephritis.

66 Norris, John (1985), p.20.

4

BRISTOL FASHION AND SHIP-SHAPE

'...THEY OUGHT NOT TO HAVE BUILT THEIR SHIPS OF SUCH LARGE DIMENSIONS!'[1]

The following chapters are not an attempt to present complete accounts of Brunel's famous ships but to try and illustrate the influences and connections with South Wales that are not always apparent. The *Great Western* was the first of his ships, a design that revolutionised ocean-going travel by proving that steam navigation across the Atlantic was a practical proposition. In this South Wales played a significant part in providing coal, the superior steam-raising quality of which allowed continuous steaming over such a distance and within the design capacity of her bunkers. Providing sustained power under such constraints was of crucial importance in order to provide a viable passenger and cargo service across the Atlantic. Other links to the *Great Western* include a major refit of the ship during her first year of operation at Pembroke Dock and the attempt made by Newport Dock Co. to attract the ship away from its Bristol base. Brunel's foray into steam navigation has been frequently told, stemming from a remark made at an early meeting of the GWR directors that the line was too long. There in October 1835 he issued, to quote Rolt, 'his most celebrated challenge', exclaiming; 'Why not make it longer, and have a steamboat go from Bristol to New York and call it the *Great Western*?'[2] Brunel was challenging criticism about the length of the main line from London to Bristol and arguing that it should be made longer, creating, in effect, a transatlantic railway on which passengers could continue their journey between Bristol and New York. Like a locomotive it would be steam driven and not at the mercy of wind and currents, allowing it to run according to published timetables.

Some thought that his idea was not in jest and took it up, which led to the establishment of the Great Western Steamship Co. (GWSC) in Bristol. Peter Maze, Bristol cotton merchant, member of the Society of Merchant Adventurers and Bristol committee member of the GWR, was to become its first chairman.[3] Captain Christopher Claxton, a half-pay naval officer who held the public post of Quay Warden at Bristol, was appointed managing director. One individual also present who possibly convinced the two last mentioned to join the enterprise, or 'experiment' as he would later refer to it, was Thomas Richard Guppy.[4] An engineer, Guppy was running his father's sugar-refining business with his brother from 1824, and in the autumn of 1832 he was one of four influential Bristol men who had resolved to press forward with the idea of a Bristol to London railway. He would therefore have first made Brunel's acquaintance around this time or earlier with Brunel becoming a family friend.[5] Guppy had a slightly different take of that *Great Western* 'eureka' moment. At the luncheon to celebrate the opening of the GWR to Maidenhead on 31 May 1838, Guppy related that the experiment was '...first proposed at a festive entertainment while watching the progress of the Great Western Railway Bill before Parliament.' He admitted that it was after dinner, following one of those occasions at the House of Commons, that the proposal:

Above: *Pembroke Dock showing the Royal Navy Dockyard. (*The Book of South Wales, the Wye, and the Coast, *by Mr and Mrs S.C. Hall, 1861)*

Left: *The* Great Western, *an aquatint published in 1840. (Courtesy of the Brunel Society)*

…was made of building the vessel which had just completed her successful voyage. [A gentleman at the other end of the table exclaimed, '*In vino veritas*'] If there was any truth in that observation, the Great Western steam-ship had given them an excellent practical illustration of it. (Cheers and laughter)[6]

Maze, Claxton, Guppy and other supporters, led by Brunel, moved quickly and during the following month Welsh newspapers such as *The Cambrian* and the *Glamorgan, Monmouth and Brecon Gazette and Merthyr Guardian* reported (in their editions of 21 November 1835) that a company had been projected to establish a steamship connection between Bristol and America. Claxton had first met Brunel in 1832 when he had been invited to survey the Floating Harbour by the Bristol Dock Co. From this association their friendship would continue throughout Brunel's lifetime, and Brunel was to described him on one occasion as a warm friend, '…changeable and very capable of being a devil of an opponent…'[7] Brunel would call upon Claxton's expertise on many occasions and in matters other than steamships; indeed, anything with a maritime dimension, from the surveying of harbours for his railways to the floating out of bridge sections. A forceful character, as Brunel describes, he was also a campaigner for naval improvements and British shipping interests. He was also prepared to risk his life to save others; it was reported that he had saved nine people from drowning and had been awarded the Royal Humane Society's medal for his eighth rescue. This was something that Brunel had in common with Claxton as he had also been awarded a Royal Humane Society medal in 1828 for saving lives during the flooding of the Thames Tunnel. In 1815 Claxton had written an account of his naval career, published as 'The Naval Monitor', and in the second (1833) edition he talks about serving in a revenue cutter, his being the first naval appointment to that class of vessel: '…her name was the *Tartar*, and while in her I caught a wife. I have since ruralised in Wales, and got children, sometimes one at a time, sometimes two…'[8]

Brunel had considered the challenges of steam navigation long before this: his father had been much engaged in this area and had played a part in promoting steam power for coastal trading. Marc Isambard considered the merits of steamships as early as 1814, and in June 1816 it was reported that he had tried out his experimental vessel on the River Thames when it was decided that '… she was the best going vessel on the river.'[9] This, the 112-ton *Regent*, went from Blackfriars Bridge to Battersea Bridge in thirty minutes and then back through London Bridge in fifty-two minutes. The great reduction in weight of her mechanical parts was remarked upon, the report stating that the 24hp steam engine, the paddle wheels '… and the machinery necessary to give and convey the movement, weighs only five tons.' It was powered by double-acting engines driving paddlewheels incorporating his designs, the engine constructed by Maudslay, Sons & Field.[10] This was regarded as a great success and greatly excited the young Brunel, but there was some opposition by older generations who were afraid of these new-fangled 'smoking monsters', particularly at Margate as when Marc Isambard arrived there by the *Regent* he was refused a bed at the York Hotel.[11] A few years later, in the 1820s, services such as the London to Gravesend run would be well established, serving the estuary towns of Ramsgate, Margate and Southend as well as providing the steamboat experiences of Charles Dickens' youth. Fortified by the success of the *Regent*, Marc Isambard tried, following much work and perseverance, to induce the Admiralty to experiment with steam vessels for towing ships out to sea. Trials were undertaken mainly at his expense, some payment was promised but eventually withdrawn, and the trials discontinued because '…the Admiralty considering the attempt "too chimerical to be seriously entertained".'[12]

A powerful image that in some ways illustrates the worst fears of the Admiralty was that painted by J.M.W. Turner (1775–1851) in 1839. Turner was a regular traveller on steamboats and frequently visited Margate by such means; his famous painting of the *Fighting Temeraire* shows the newcomer, steam, and Nelson's navy. The ghostly outline of the Trafalgar veteran is seen being towed to the breakers yard by one of these 'smoking monsters', predicting the inevitability of steam power and the reluctance of the Admiralty to change. Despite the pioneering work of his father, many writers, including Rolt, are of the opinion that Marc Isambard did not feel that trans-oceanic navigation by steam power was possible. He refused to act as a consulting engineer to a steamship company operating between Britain and the West Indies in about the year 1824, saying that '… steam cannot do for distant navigation, I cannot take part in any scheme.'[13] A contemporary, George Thomas Clark, an engineer who had worked under Brunel on the GWR and later became the resident trustee of the Dowlais Iron Works claimed to have known Marc Isambard well, at one time meeting him almost daily. He confessed in a letter to *The Times* that '…I also, at first, did not believe that he ever could have expressed himself so decidedly against Transatlantic communication by steam; …'[14] He does concede that such a remark could have been made with regard to the 1824 scheme (adding that he would like to see the whole letter from which the quote was taken), but not that he used the word 'impossible' (which was used by Marc Isambard's 1862 biographer, Richard Beamish, in the same context as above):

> However, this may be, certain it is that before 1830, Brunel [Marc Isambard] had completely changed his opinion, and was an ardent promoter of and believer in the scheme brought forward and successful in the hands of his son. I trust I shall not be thought hypercritical if I remark that Brunel does not use the word 'impossible' attributed to him. It was a word seldom used by him or his son. They preferred the word 'impracticable'.

Clark also objects to the coupling together of the change of opinion of Marc Isambard and that of an individual whose name is frequently raised in this episode, that of the eccentric scientist Dr Dionysius Lardner (1793–1859). Lardner had criticised Brunel's Box Tunnel during the parliamentary proceedings of the GWR, with the claim that runaway trains entering the tunnel would emerge at 120mph, which would suffocate all the passengers! Lardner had not included factors such as friction and air pressure and Brunel patiently pointed out they would prevent the train from exceeding 56mph in such circumstances. Following a lecture at Liverpool in December 1835, he is recorded as stating that a voyage from New York to Liverpool was perfectly chimerical, '…and they might as well talk of making a voyage from New York or Liverpool to the moon.'[15] In August 1836 the British Association for the Advancement of Science met at Bristol, and Lardner was to talk on the subject of 'Transatlantic Steam Navigation'. Again he argued that steam navigation across the Atlantic was not possible, the basis of his argument being that the power of the engines must be increased with the size of the vessel; doubling the size of the ship meant a doubling of the engine power. Brunel was in the audience and as before he was to expose several errors in his calculations. Unfortunately the details of this exchange were not recorded, but Brunel put forward his argument that '… while the tonnage of a ship is increased as the cube of her dimensions, the resistance is increased only about as the square.'[16] Brunel was to prove his argument by practical means with the *Great Western* steaming across the Atlantic. As that day approached, Lardner increasingly modified his stance on the subject. In September 1837 he again addressed the British Association for the Advancement of Science, this time meeting at Liverpool, but was now not denying that such a voyage might be practicable, only that he did not believe it would be profitable.[17]

Lardner's change of opinion on high speeds on railways and the use of the Blackwall rope are also cited by Clark, expressing an opinion that '…Lardner was a clear expounder and a clever mathematician; but he was no engineer…' Brunel was an engineer but he was not a shipbuilder, so he suggested the setting up of a 'building committee' which would guide the construction of the *Great Western* whilst William Patterson would superintend the building of the ship. The committee consisted of Brunel, Guppy and Claxton.[18] Patterson was an established shipbuilder, trading as Patterson & Mercer, at a shipyard near the Princes Street Bridge at Wapping, Bristol. Early in June 1836 the largest keel in the world was laid and work on the largest steamship yet to be constructed got under way. The *Great Western* measured 236ft in length with a displacement of 2,300 tons, and her engines, built by Maudslay, Sons & Field of London, developed 750 indicated-hp for the 28ft-diameter paddle wheels. Patterson's work could be said to be 'ship-shape and Bristol fashion' an expression that first appears in print around 1840. It appears as 'Bristol Fashion and Ship-Shape' in Admiral W.H. Smyth's book, which states that it was said when Bristol was in '… its palmy commercial days, unannoyed by Liverpool, and its shipping was all in proper good order.'[19] The launch took place on 19 July 1837 to a great gathering of crowds. The *Bristol Mirror* told its readers that a great shout of 'She moves' went up when the ship glided into the water, and that:

William Patterson (1795–1869), an experienced shipbuilder who turned Brunel's ideas for transatlantic steam navigation into a reality. (Courtesy of the Brunel Society)

> As she left the shore, Lieutenant Claxton performed the usual ceremony of dashing a demijohn of Madeira upon the figurehead of Neptune at the bows, and she was named by Mrs. Miles, who also cracked her bottle against the side, the *Great Western*. At the calculated distance she was checked by a chain cable, and brought up within a few feet of the opposite shore, without the smallest accident…[20]

Claxton would also refute Lardner's statements regarding the impossibility of transatlantic steam navigation with calculations based on the steam-raising power of coal from the coalfields of south-west Wales, and this is dealt with below. Another Welsh product with the highest reputation was chain cable supplied by Brown Lenox of Pontypridd, but the supplier of the chain cable used on the *Great Western* is not known to the author. The total cost of 'Rigging, sails, anchors, cables etc' for the *Great Western* is given as £1,452 16s 5d.[21] It is likely that the chain cable was supplied by Acramans of Bristol. William Edward Acraman was one of the directors of the Great Western Steam Ship Co. and his company forged both paddle shafts for the ship. The company manufactured under licence Captain Samuel Brown's patent chain cable, and Acraman's would supply the *Great Britain* steamship.[22] After Brunel's death there was some correspondence in *The Times*, started by William Patterson junior, that Brunel was not the ship's designer. Claxton replied by saying that it is beyond dispute that the lines of the ship were designed by Patterson but that both the *Great Western* and *Great Britain* were '…emanations of Mr. Brunel's genius.' He went on to say:

> Mr. Patterson drew the lines; Mr. Brunel, Mr. Guppy, and myself, often sat over them; Mr. Patterson got instructions and made his own calculations accurately; Mr. Brunel made his often by my side… That as a shipbuilder [his role] is established, and I have always thought, and still believe, he was proud in being called upon to assist Mr. Brunel.[23]

The *Great Western* left Bristol on 8 April 1838 and arrived at New York on 23 April 1838, a voyage of fifteen days and ten hours. It was originally intended to sail on the 7th but Captain Hosken decided that because of delays in loading stores and cargo, which was hampered by the weather, the ship would lay at her moorings until the next morning. The ship had been at these moorings since 2 April following her return from London where she was fitted out, accommodation completed and her engines, boilers and machinery installed by Maudslay, Sons & Field. This was the voyage that had resulted in Brunel taking a forced recuperation on Canvey Island, and news of the fire, with some reports claiming the *Great Western* was a total loss, put a number of passengers off travelling on her first voyage to New York.[24] The ship set out on her maiden voyage, which turned out to be a 'race' with the *Sirius*, a story too well told to be repeated here. The return of the *Great Western* to Bristol was accomplished in exactly fourteen days and the return trip of her second Atlantic voyage in June of that year was accomplished in twelve days and fourteen hours. On her fifth voyage from New York to Bristol, arriving on 7 December 1838, she had steamed 2,835 miles in thirteen days and six hours, an average of 218 miles per day. The greatest progress made in any one day was 260 miles, and the least 176.[25] After this return and before the end of 1838 the *Great Western* would sail to what was listed as 'drydock for check and overhaul', but it was to be much more significant than that. Unusually the work would be carried out in Her Majesty's Royal Dockyard at Pater, or Pembroke Dock, and permission had to be obtained from the Admiralty. Denis Griffiths writes that the ship needed little in the way of repairs at Pembroke Dock but that major new deck works would be carried out, removing the temporary cabins put up in the time spent at Kingroad between the second and third voyages.[26] Cargo capacity would be increased by an extra 50 tons as a result of this work.

The *Great Western* was to sail from Bristol to Milford Haven on 20 December 1838 to spend time in Her Majesty's Dockyard at Pembroke, the first two weeks or so being spent outside the dock in Milford Haven, and she returned on 17 January 1839.[27] The press announced that the ship had arrived in the Haven that day but she was to stay outside the dockyard and lay at anchor over the Christmas and New Year, a delay blamed on the state of tides, and it was not until after 3 January 1839 that the ship could be hauled into Pembroke Dock for her refit. The directors of the Great Western Steam Navigation Co. had issued a notice concerning the delays because of the impact on her sailing schedule.[28] *The Cambrian* newspaper only appears to have picked up on her presence there, following this, in its edition of 5 January 1839, but she then featured every week following in the newspaper until 26 January. The story published on 8 January 1839 concluded by saying that she needed considerable repairs. When the ship did get into the Royal Dockyard she did undergo considerable work. A new poop, for the convenience of passengers, '…which is to be splendidly fitted up with several new berths…', was carried out by the artificers at the dockyard.[29] *The Ipswich Journal* carried a full report on the work carried out at the Royal Dockyard:

> During her stay in the Royal Dock-yard, great alterations have been made on board, all of which we are assured, will very materially contribute to the comfort and convenience of the passengers. The whole of the lower berths under the saloon have been thrown into a cargo space, and the passengers' berths reduced to a number which must ensure every comfort and accommodation. The house on deck has been removed, and in lieu thereof the cuddy has been carried forward more to the middle part of the vessel, and has a range of cabins on either side. The splendid saloon is lighted from upper skylights. Great additional room has been gained by the alterations, and a space under cover is obtained sufficient for a promenade for the whole number of passengers. The whole of the cooking establishment has been very much increased, and several other conve-

niences have been built on deck. The vessel, on examination was found to be in the very best order, without even a single strain or so much as a ruck in her copper.[30]

It was also stated in *The Ipswich Journal* that the alterations had been undertaken at the suggestion of Captain Claxton, RN (the managing director), '…a gentleman whose scientific attainments have been largely shown in the construction of this magnificent vessel.' The commander of the *Great Western*, Captain Hosken, had stayed with the ship while she was at Milford Haven and had advocated such alterations '… as from experience he knows will add to the comforts of the passengers.' An impression of the dockyard at this time can be gleaned from a visit to the dockyard made by the Revd Joseph Romilly in August 1840, who noted that several 80-gun ships were being built there. He and his party went into the 'skeleton' of the *Centurion* which would be launched four years later. When they visited the model room they had '… the plans and drawings of ships explained by a very intelligent man.' They also went onboard a 26-gun frigate, HMS *Iris*, which had been launched two weeks before.[31] An idea of travelling times by ordinary steam packets is also given by the Revd Romilly when he and his party travelled from Hobbs Point on Milford Haven to Tenby; the ten miles took two and a half hours. They then boarded the steamboat *Star* at 1.30 that afternoon and arrived twelve hours later at Clifton, where Romilly had to jump from the paddlebox of the *Star* onto another to get to the landing stage.[32] The return time for the *Great Western* over a slightly shorter distance is given by the last entry in *The Cambrian* on the *Great Western*'s progress in Pembrokeshire. It reported on the extraordinary speed of the *Great Western* on its return to Kingroad, Bristol, a journey accomplished in the remarkably short time of nine hours (including an hour and a half's delay adjusting her machinery). Another newspaper stated that she left Milford at 10.00 a.m. on Saturday, lay-to for a full hour and a half off St Gowan's Head, and arrived at her moorings at 7.00 p.m. of the same day. It went on to say that the distance from Tenby to Kingroad was reckoned to be ninety-four miles, in which the fastest time for packets on the station was eight hours: 'The Great Western on this occasion performed it in six hours!!!'[33]

Her first voyage in 1839, her sixth Atlantic voyage, would be on 28 January, and it was reported that nearly all her berths were taken. Despite all these improvements and the success of the venture, the *Great Western* was not able use the facilities of Bristol Docks due to the problems of high charges and the fact that the *Great Western* would have to remove the lower part of her paddle wheels to get through the locks at Cumberland Basin and into Bristol Docks. The company continued to use moorings at Kingroad some seven miles downstream from Bristol, situated off Broad Pill. Early in 1839 this situation and the underlying reasons behind it was raised by *The Bristol Mercury* in a story headed 'The Rival Docks and the "Great Western"'.[34] The newspaper was concerned that the high charges levied by the Bristol Dock Co. was driving away trade with excessive tonnage dues which, it claimed, were 300 per cent over and above that of London and Liverpool. The problems of the insufficient width of the entrance locks were also highlighted, and that the response by the dock company to shipowners who complained was that: '…they ought not to have built their ships of such large dimensions!' The newspaper believed that, unless forced to do so, they would not even adopt measures to accommodate the smaller classes of steamers at the entrance to the Cumberland basin, '…to say nothing of the improvement demanded by vessels of the size of the *Great Western*'. The *Bristol Mercury* felt it had to once again raise these issues at this time because of a new threat to the ship abandoning Bristol as its home port being posed by new dock works on the other side of the Bristol Channel at Newport. How many vessels would follow the example of the *Great Western* and leave Bristol?

The dock works were actually Newport's first dock, the Old or Town Dock, of which work began following the passing of the Newport Dock Act in 1835. *The Bristol Mercury* referred to a report from the *Monmouthshire Merlin* concerning a meeting in which the Newport Dock proprietors convened to consider adapting the entrance lock for the admission of steam vessels rather than the size hitherto proposed. Some parties objected to the additional cost in deviating from the original plan, stated to be £6,000 (including iron gates), and called for the opinion of their consulting engineer, William, later Sir William, Cubitt (1785–1861), to be taken seriously. A resolution was finally carried by a large majority. This was that: '…it is desirable that the dock be constructed upon a scale to admit vessels of the dimensions of the *Great Western*…' subject to their complete satisfaction with the plans and estimates of the proposed enlargement. If the contemplated improvements at Newport were carried into effect, ensuring completion probably in the spring of 1840, this would be a great incentive for the *Great Western* to avoid charges of up to £100 per voyage by continuing to operate from Bristol:

> It is against all experience to suppose that the proprietors of this splendid ship, 'our pride', as the Editor of the *Mirror* calls her, will continue to incur the risk and expense consequent on her laying at Kingroad, when a convenient berth is provided for her at Newport, and especially as they will then no longer be subject to the Bristol port charges, which are now, without a shadow of justice, demanded of them.

The *Great Western* would return again to Milford Haven on 4 April 1843, spending thirteen days there before returning to Liverpool. She presumably received her drydock inspection in the new dock at Pembroke, work on which included an extension of the yard to the west, as announced in August 1839.[35] In the meantime the first Bute Dock was to open at Cardiff, described as the stupendous works undertaken by the Marquis of Bute, and taking place with much celebration on 9 October 1839. The entrance lock was 36ft wide and therefore also not capable of allowing the *Great Western* through, with its breadth over paddle wheels at 59ft 8in and a very tight fit with its paddle wheels removed (35ft 4in). At the celebration dinner that night at the Cardiff Arms Hotel, the man who had persuaded Brunel to build the Taff Vale Railway, Sir Josiah John Guest, spoke about the natural advantages of the port. He was con-

Newport Town Dock opened on 10 October 1842 boasting 64ft-wide lock gates, an attempt to attract the Great Western to Newport. Photographed in 2008, only the silted-up lock entrance remains, the dock having been in-filled. (SKJ photograph)

vinced that a large amount of trade would come to Cardiff as a result of the new docks, and he could not help referring to the old rival of South Wales:

> Bristol had been termed the Queen of South Wales (a laugh); this however, was no longer the case, the time, he thought, was not far distant when a considerable part of the trade which now wound its way up the sluggish and tortuous course of the Avon would centre in Cardiff (loud and prolonged cheers).[36]

It was not until 1842, however, that the last of several contractors working on the dock, Messrs Rennie, Logan & Co., completed the work which, with warehouses etc., covered 24 acres and was opened on Monday 10 October 1842 in the presence of at least 25,000 spectators.[37] The engineer was James Green, who had also designed the Bute Dock at Cardiff, and his dock at Newport would cost £150,000 to build. Occupying an area of 4½ acres, the basin was 795ft long and 240ft wide, capable, it was said, of '...affording accommodation to 50 vessels of large class at one and the same time.' Rennies were also the contractors for the gates which were 64ft wide, then the largest in the world in a lock entrance, being 220ft long with a depth of 35ft 6in. The gates were built with iron transoms which were opened by machinery, taking about twelve minutes to open the gate into the basin and a shorter time for the gate from the river. The width of the lock entrance was 61ft, giving a tight squeeze if the *Great Western* had ever ventured into the dock. However, on opening day the first to enter the dock at 10.20 a.m. was the *Usk* steamer towing Thomas Powell's schooner the *Henry* behind, and this was followed by the Channel mission ship the *Eirene*, then the steam tug *Hercules*, '...having in tow a fine vessel called the *Great Britain*...' this was not of course Brunel's ship but an 800-ton Newport-registered barque owned by Joseph Latch.[38]

The width of the lock entrance did not capture the *Great Western* in terms of using the dock, but some years later, in 1882, it did 'capture' two vessels using the lock. It had become common practice to admit two ships at a time through the lock, but on this occasion the vessels got jammed, the gates could not be closed and the water escaped with the tide.[39] It was claimed at the opening ceremony that Newport was now an infant Liverpool, and by 1846 coal being shipped through the dock and the existing wharves was some 130,000 tons more than Cardiff.[40] This situation did not last long as the high-level branch of the TVR at Cardiff came into use in 1848, the second Bute Dock opened in 1855 and although Newport responded with an extension increasing the water area of the Town Dock to 12 acres, it would not outstrip Cardiff again.[41] The MP Richard Blakemore took the chair at the dinner that evening, and his remarks would follow in a similar vein to those made by Sir John Guest at the opening of the Bute Dock three years earlier, Blakemore referring to Bristol's 'palmy commercial days' and that it was within his own memory that Bristol possessed all the carrying trade of the interior, that all the products of Worcestershire, Staffordshire, Shropshire etc. would be conveyed through her port to the world. He asked what she does now? – giving the answer that she does not carry an ounce: 'Liverpool, by the improvements she made in the River Mersey, has taken away this trade from Bristol. For the want of using similar exertions, Bristol has lost the carrying trade.'[42]

However serious the Newport offer may have been, it does not appear to have been taken seriously, with the GWSC deciding to use Liverpool earlier that year, the company planning to alternate between Liverpool and Bristol, with the first arrival at Liverpool from New York taking place on 11 May 1842.[43] A major disadvantage was that Newport was not yet connected by railway. The SWR would not be open until 1850. In the years that the *Great Western* was still using Bristol her best outward time was fourteen days and one and a half hours, and her best return was accomplished in twelve days and ten hours.[44] This alternation of 'home' ports continued until the beginning of 1843, after which Bristol was only for overwintering and overhauls. She left Bristol for the last time on 29 April 1847 under new owners, the Royal Mail Steam Packet Co.[45] Apart from a dock that was wide enough to accommodate her, there was another good reason why the *Great Western* might use Newport as a base: to be closer to a ready supply of coal. Newport was the natural outlet for the coal most favoured for use in the ship; Tredegar coal. It was not the original coal of choice, as in 1836 Claxton had singled out Llangennech or Graigola coal as being the most suitable for steaming purposes in his public letter when he argued against Lardner's statements on steam navigation and the figures used for coal consumption.[46] Lardner had stated that in the region of 1,400 tons would be required for engines of 400hp to run the distance to New York. Claxton believed that 5½lb of Llangennech coal was equal to 9lb of other sorts, so that Welsh coal would give a fuel efficiency of 40 per cent over other coals and it was therefore possible to carry enough coal for the voyage. Interestingly, both Llangennech and Graigola had connections with railways built or surveyed by Brunel's resident engineer on the TVR; George Bush.[47] In 1840 Bush surveyed the Swansea valley for a proposed railway that would provide a connection to the Graigola collieries, but was to die before this was taken any further. Before his work on the TVR, Bush had been engineer of the Llanelly Railway which provided a railway connection and a dock for the export of Llangennech coal. Samuel Lewis reported that more than 100 tons of coal and culm were annually exported from the dock at Llanelli, and that '…some of the coal owing to its peculiarly fine quality, [was] being shipped to France, and to the Mediterranean, for the use of the steam-boats.'[48]

The Llangennech Coal Co. was financed by a partnership of London merchants, who sent its first consignment of coal to London in 1824 from the St David's colliery near Llanelli, which at the time was the deepest mine in Wales. This consignment was presumably sent from its own quay on the River Loughor, along which barges of up to 200 tons burden would be towed to Llanelli for onward shipping of this 'very superior quality' coal which '…obtains a high price in the London market…'[49] Despite the company having an improved outlet for its coal with the opening of the Llanelly Railway, it along with other river users no doubt maintained the right to river navigation on the Loughor for which a swing bridge was incorporated into Brunel's SWR viaduct. The western coalfield, including Llangennech, Graigola and the colliery originally started by General Warde, enjoyed the lion's share of the Welsh coal sent to London at this time. Typically, as in 1832, the Admiralty would advertise for tenders for the supply of Llangennech coal, a description that could be met by coal from that district as in the case of 'Nevill's Llangennech', i.e. coal mined by R.J. Nevill in the Llanelli area.[50] The coal was required for use by naval steam packets, with costs for shipping to ports such as Gibraltar and Malta, with other tenders often specifically requesting Bryndewi (Neath), Graigola and Llanelly coals. Between 1840 and 1844, five out of eight Admiralty contracts were supplied by collieries in the western coalfield, but increasingly the eastern valleys also began to win steamship contracts.[51] In 1845 Joseph Hume, then the MP for Aberdeen, wrote to the Admiralty questioning the price and the choice of coal purchased for its steamships. The Admiralty responded by commissioning tests by Sir Henry de la Beche (1798–1855) and Dr Lyon Playfair, with thirty-seven samples from Wales and seventeen from Newcastle. These tests found that the best Welsh coals lit easily, blew steam up rapidly, produced a fine, clear fire with no clinker and gave off very little smoke. The 1848 report established South Wales's coal as the premium coal, not just in Britain, but throughout the world.

There would be counter-claims and further reports but nothing could stop the growth of the South Wales coalfield in satisfying not just for the navy but industrial demand worldwide.

The position of major coal exporter within the South Wales coalfield would, however, change with the commercial exploitation of coal seams in the western coalfields of Glamorgan and Monmouthshire, particularly Aberdare steam coal, by the construction of railways and docks in the 1840s.[52] Brunel, of course, contributed to this development as the engineer of the TVR, a railway whose prosperity grew enormously with the opening up of the mineral resources of the valleys, and in choosing Rhondda coal as the main source of coke used by locomotives of the GWR. In 1845 GWR locomotives consumed 21,919 tons of coke whilst travelling a distance of 1,240,412 miles. Those figures were actually collated by Lardner for his book *Railway Economy: A Treatise on the New Art of Transport.*[53] Lardner was a prolific author, and being proved wrong on a number of occasions when up against the Brunel camp does not appear to have affected his output. The Royal Mail Steam Packet Co. and the Peninsular & Oriental Co. obtained their coal from the Risca colliery of John Russell & Co. The Great Western Steam Ship Co. also decided on using coal from this side of the South Wales coalfield and, after extensive trials, decided in favour of coal supplied by Samuel Homfray's Tredegar Iron Co. Homfray, the vice-chairman of proceedings at the opening of Newport Dock, took the opportunity to state that the dues at Newport Dock would be lower than any other port in the kingdom. This would, of course, make coal exported by his trading subsidiary, the Tredegar Coal Co. at Newport, to ports such as Bristol a cheaper option.[54]

The choice of coal was a prudent commercial decision and one that proved to be correct on several occasions. Claxton had issued a statement in a letter to the *Mechanics Magazine* concerning the suitability of different coals burnt on the ship during the homeward passage in July 1838. Canadian coal, from the Nova Scotia mines of Pictou, had been bought in at New York to try and reduce the high bunker costs of Tredegar coal. It proved to be completely unsuitable, Claxton confirmed: 'Of the four descriptions of coals used on this voyage, the best being that from Tredegar.'[55] The *Great Western* had to use coal shipped from Liverpool on a return voyage from New York in December 1838 which burned so poorly that difficulty was experienced in keeping up steam. It was found that less steam was produced with 35 to 40 tons of these Liverpool coals than with 25 tons obtained from the Tredegar Coal Co. Towards the end of that same year, and a week before the *Great Western* set sail for Pembroke Dockyard, Claxton was writing to the Admiralty regarding the invitation to tender for a steamer service of twelve round-voyages annually between Britain and North America, that had appeared in the press in November 1838. To say that this mail contract was something of a rush job, being advertised with bids required by 15 December, for a service to start by 1 April 1839, was an understatement, and perhaps the intended starting date was appropriate. It was also rather perverse as no one company had sufficient ships for the job and there was not enough time to construct vessels, particularly for a one-year contract, which is what was originally proposed. Claxton wrote his letter setting out the position of the GWSC, two days before the bid deadline, in which he gave his opinion on transatlantic steam navigation, asked for more time, a change in the specification given for the ships and a longer contract period.

It was fair to say that the GWSC had something of a monopoly on transatlantic steam navigation at this time. The ship had made five round-trip crossings, and probably Claxton felt they could rewrite the conditions of the Admiralty mail tender. However, the Admiralty did not like being told about steamship matters and how the service should be run, particularly by a half-pay naval officer, and rejected his proposal on 10 January 1839.[56] The GWSC's case would not have been helped by the criticism made by Guppy in the autumn of 1837, no doubt still fresh in their minds, of the steamship designs being adopted by the Royal Navy. There were only two responses to the tender invitation, the other coming from the owners of the *Sirius*; the St George Steam Packet Co. Although technically with two ships (both of which had been built

Bedwellty House, Tredegar, photographed on 24 January 2008. The house was rebuilt in 1825 by Samuel Homfray of Penydarren whose Tredegar Iron Co. was to supply coal to the Great Western *steamship. (SKJ photograph)*

for trade around the British Isles), they could have started a service by the deadline but bigger and better ships would be needed to maintain the service. Their tender was also declined, the bigger deadline passed and the Admiralty had no bids they were prepared to accept. In this situation the Admiralty found themselves listening to the proposals of a largely unknown Canadian businessman called Samuel Cunard (1787–1865), some two months after the tender deadline, on 11 February 1839. The contract with the British & North American Royal Mail Steam Packet Co. (but to be known simply as the 'Cunard Line' from its inception) was signed on 4 May 1839, and *Britannia*, the first ship of the line, was launched on 5 February 1840. Her maiden voyage on 4 July 1840, later than planned and for which special dispensation had to be sought from the Admiralty, was from Liverpool. In January 1842 Dickens booked the *Britannia* for a passage to Boston but the voyage did not live up to his expectations.[57]

Not winning the contract was a setback as it involved the loss of financial subsidy and the greater commercial viability that the mail tender would have brought to the GWSC. This appears somewhat unfair as the GWSC had taken all of the speculative risks in proving the viability of steam navigation whilst Cunard now reaped the benefits. Cunard's guiding philosophy was one of conservatism as it would not adopt any innovations or advances in steam navigation until they were proved by other steamships and their operators. The new line did not have everything its own way, however; it certainly took some of the passenger traffic on the Atlantic run away from the *Great Western* but the losses were not damaging as passengers on the rival service tended to be going to or from Nova Scotia or Boston.[58] All mail, unless specifically marked to go by the *Great Western*, now had to go via Cunard and anyone in the service of the British Government was obliged to use the mail steamers as the contract allowed for reduced rates for such passengers. Whenever the two services were in direct competition with similar sailing times, the *Great Western* came off best, so much so that Cunard fares were reduced to try and attract customers away from the loyal following that Brunel's ship enjoyed.[59] The *Great Western* set the standard for transatlantic steam navigation and built a reputation for safe, fast, reliable and regular sailings. It maintained a healthy share of the Atlantic passenger traffic and remained profitable even without a mail contract subsidy. The successful history of the ship, from conception to the end of her career in 1857, was based on many factors, and the part played by Welsh coal cannot be overlooked.[60] This concludes this brief account

Welsh Back, Bristol. Bristol was not only a major seaport, described by Daniel Defoe in 1725 as '…the greatest, the richest, and the best port of trade in Great Britain, London only excepted', but also the hub of a vast coastal trading network. Part of the wharf area below Bristol Bridge was named Welsh Back because you could always find a 'boat back to Wales', and 'back' is said to be derived from 'bac' or 'bec' – the old English word for ferry. (SKJ photograph)

Brunel would, no doubt, have been familiar with the establishment and Bristol landmark that was the Llandoger Trow, an inn dating back to the seventeenth century, built by a Captain Hawkins. It was named after his flat-bottomed sailing barge, or trow, that sailed the Severn and went up the River Wye to Llandogo. Such barges forced Brunel to come up with the novel design for his Wye Bridge at Chepstow, incorporating the superstructure above the rail deck and allowing a 50ft headway at high tide. (SKJ photograph)

of the *Great Western*, a successful ship in her own right but one that was limited in terms of further innovation. The *Great Western* represented a great leap forward but the advancement of ocean-going steam navigation would be limited if it was based on wooden paddle ships. Brunel was one of the few who realised this and again he would take the lead in the future of steamship design when considering the construction of further ships to follow the *Great Western* for the GWSC. In the end only one ship would follow in this particular story but Welsh coal would also play an important part, particularly at what had seemed to be the end of that ship's working life.

CHAPTER 4 NOTES

1 *The Bristol Mercury*, 26 January 1839; a story headed 'The Rival Docks and the "Great Western"'.

2 Rolt, L.T.C. (1959) p.249, *Isambard Kingdom Brunel* (Longmans, Green & Co.: London). This story appears in the biography by Brunel's son Isambard in 1870 (p.233), *The Life of Isambard Kingdom Brunel* (Longmans, Green & Co.: London, 1870, reprinted by David & Charles, Newton Abbot, 1971).

3 His Bristol cotton mill of 1838, which opened in Barton Mill, was called the 'Great Western Cotton Mill'.

4 Samuel Guppy was a prominent Bristol merchant with an interest in the copper trade, and in 1815 his son undertook his own nine-year-long educational tour of Europe, learning engineering, architecture and painting on the way, before returning to Bristol in 1824.

5 Buchanan, R.A. (2002), p.200, *Brunel, The Life and Times of Isambard Kingdom Brunel* (Hambledon & London: London). If it were not for the conventions of the day, his sister Sarah Maria Guppy would also have made her way as an engineer, having several inventions to her name.

6 *The Times*, 2 June 1838. The *Great Western* returned to Bristol from her first Atlantic crossing on 22 May 1838. On the return trip that day, after luncheon, Guppy walked along the top of the railway carriages.

7 Buchanan, R.A. (2002), p.58 (private diary entry for 30 August 1833, p.90).

8 Claxton, Captain Christopher (1833), p.4, *The Naval Monitor*. First edition published in 1815.

9 *The Liverpool Mercury*, 5 July 1816 (reporting from an edition of the *Morning Chronicle* that had reported her run on the Thames that previous Saturday morning).

10 Clements, Paul (1970), p.61, *Marc Isambard Brunel* (Longmans, Green & Co.: London). Maudslay's role is confirmed by the *Liverpool Mercury* report of 5 July 1816.

11 Young, G.M. (1934), p.395, *Early Victorian England*, Vol.1, 1830–65, Chapter 'The Mercantile Marine' by Basil Lubbock (Oxford University Press – Humphrey Milford: London). Margate had good landing facilities, unlike Southend whose development as a resort was hampered until a pier was built.

12 From the obituary notice of Sir Marc Isambard Brunel (1769–1849) in Minutes of Proceedings of the Institution of Civil Engineers, Vol.10, Session 1850–51, pp.78–81.

13 Rolt, L.T.C. (1959), p.248.

14 *The Times*, 31 May 1879. Written by Clark from his and Brunel's old club, the Athenaeum, on 28 May. Marc Isambard's opinion on steam navigation was made at least eight years before the problem was solved whereas Lardner's was on the very eve of its solution. For Clark see Vols I and II.

15 Griffiths, Denis (1985), p.11, *Brunel's 'Great Western'* (Patrick Stephens: Wellingborough).

16 Brunel, Isambard (1870, reprinted 1971), pp.240–1, *The Life of Isambard Kingdom Brunel* (Longmans, Green & Co.: London, 1870, reprinted by David & Charles, Newton Abbot, 1971).

17 Brunel, Isambard (1870, reprinted 1971), p.240.

18 Griffiths, Denis (1985), p.15.

19 Smyth, Admiral W.H. (revised by Vice-Admiral Sir E. Belcher and published in 1867), *The Sailor's Word-Book: An Alphabetical Digest of Nautical Terms* (Blackie & Son: London). Bristol fashion was added later and is first seen in print during Bristol's heyday as a trading port, in Richard Dana Jr's *Two Years before the Mast*, 1840: 'Everything on board's ship-shape and Bristol fashion'.

20 Griffiths, Denis (1985), p.19. Report in the *Bristol Mirror* for 22 July 1837. An illustration of the launch appears in Vol.I of *Brunel in South Wales*, p.124.

21 Griffiths, Denis (1985), p.149. Appendix 5: Financial Aspects of the Great Western Steamship.

22 Other companies manufacturing Brown's chain cable under license included Crawford Logan of Liverpool and Hawkes, Crawshay & Sons of Gateshead. The latter associated with the Crawshay family at Cyfarthfa.

23 *The Times*, Patterson's letter 21 September 1859, Claxton's reply 22 September 1859.

24 Griffiths, Denis (1985), pp.30–31. Some fifty people cancelled their bookings leaving just seven passengers on the maiden voyage.

25 *The Newcastle Courant*, 21 December 1838.

26 Griffiths, Denis (1985), p.45. The temporary measures increased the nominal first class accomodation from 128 to 140 berths, but the permanent work at Pembroke Dock would reduce it to 109, see pp.48–49.

27 The superintendent of Pembroke Dockyard at the time was Captain Samuel Jackson, C.B. (1775–1845). Previously the commander of HMS *Bellerophon*, Jackson was in post from 19 February 1838 to 23 November 1841.

28 *The Times*, 3 January 1839. In the light of the notice issued by the company, the *North Wales Chronicle* appears to have been slightly out with its article saying that the *Great Western* was taken into dock at Pembroke Yard on; '…Tuesday 1st instant, and workmen are busily engaged…'

29 *Jackson's Oxford Journal*, 26 January 1839.

30 *The Ipswich Journal*, 26 January 1839.

31 Morris, M.G.M. (1998), p.91, *Romilly's Visits to Wales 1827–1854* (Gomer Press: Llandysul). The *Centurion*, now steampowered, was part of the Channel Fleet at Milford Haven when the *Great Eastern* returned from her first Atlantic crossing and *Iris*, like the *Great Eastern*, was later pressed into service as a cable vessel.

32 Morris, M.G.M. (1998), p.91, *Romilly's Visits to Wales 1827–1854* (Gomer Press: Llandysul). The *Star* also sailed from Bristol to Milford and Haverfordwest on Wednesdays.

33 *The Ipswich Journal*, 26 January 1839.

34 *The Bristol Mercury*, 26 January 1839.

35 *The Morning Chronicle*, 12 August 1839. 'Pembroke Dockyard is to be considerably enlarged westward, and a new dock formed, agreeably to the instructions of government.'

36 *The Bristol Mercury*, 12 October 1839.

37 *The Bristol Mercury*, 15 October 1842.

38 Davis, Haydn (1998), p.129, *The History of the Borough of Newport, from Swamp to Super-town* (Pennyfarthing Press: Newport). This *Great Britain* was to be lost on 30 March 1843 in a mid-Atlantic storm. *The Bristol Mercury*, in a story published on 10 June 1843, and some weeks after her reported loss, states that the ship was of 404 tons burden, frigate rigged and built at Quebec in 1839.

39 Chapman, W.G. (1927, reprinted 1971), p.32, *'Twixt Rail and Sea: A Book of Docks, Seaports, and Shipping* (Patrick Stephens Ltd: London). Following this incident the dock company sold out to the rival Newport Dock operator; the Alexandra (Newport & South Wales) Docks and Railway.

40 Although the TVR had been completed to Cardiff it was not until May 1846 that an agreement was signed between Lord Bute and the TVR became law and facilities were put in hand to ship from the Bute Dock. See Vol.I, pp.174–175.

41 Haydn Davis states that when the Newport Dock extension opened on 2 March 1858 the opening ceremony was 'graced by Isambard Kingdom Brunel's famous ocean liner "The Great Britain".'

(Davis, Haydn, p.130 (1998). However, at this time the *Great Britain* was returning from a voyage to Bombay, arriving at Liverpool on 10 April 1858.

42 *The Bristol Mercury*, 15 October 1842.

43 Griffiths, Denis (1985), pp.89–90. The directors were forced to sell the *Great Western* in April 1847 due to financial circumstances following the refloating of the *Great Britain*. The rival Cunard ship, *Britannia*, arrived at Liverpool four days later.

44 Griffiths, Denis (1985), pp.141–146, Appendix 3, 'The voyages of the paddle steamship Great Western'. Outward voyages of less than fourteen days were accomplished from 1843 on the Liverpool to New York run.

45 Griffiths, Denis (1985), p.124.

46 *The Bristol Mercury*, 10 September 1836.

47 See Vols I and II.

48 Lewis, Samuel (1840 second edition), *A Topographical History of Wales, Vol. II*, see entries for Llanelly and Llangennech.

49 Lewis, Samuel (1840 second edition), entry for Llangennech.

50 Morris, J.H., and Williams, L.J. (1958), pp.22–3, *The South Wales Coal Industry 1841–1875* (University of Wales Press: Cardiff). His London agent being instructed to submit a tender for 'Nevill's Llangennech'.

51 Morris, J.H., and Williams, L.J. (1958), p.28.

52 Morris, J.H., and Williams, L.J. (1958), p.20.

53 Lardner, Dionysius (1850, reprinted 1968), p.81, *Railway Economy, A Treatise on the New Art of Transport* (David & Charles (Publishers) Ltd: Newton Abbot).

54 Morris, J.H., and Williams, L.J. (1958), p.84. Newport had an advantage prior to 1831 when all coal shipped there could be delivered duty free to all ports eastward of the Holms Islands in the Bristol Channel.

55 Griffiths, Denis (1985), p.51.

56 Fox, Stephen (2003), pp.85–6, *The Ocean Railway; Isambard Kingdom Brunel, Samuel Cunard and the Revolutionary World of the Great Atlantic Steamships* (Harper Perennial: London).

57 Published in his travel book, *American Notes*, his account paints a grim account of the realities of mid-winter transatlantic travel at the beginning of the steamship era.

58 Griffiths, Denis (1985), pp.54–5.

59 Although she only catered for first class passengers, the *Great Western* offered a higher level of comfort than the first class accommodation offered by her rivals.

60 The final end of the *Great Western*, witnessed by Brunel himself, has been told in Chapter 1.

5

PORTFOLIO

'WRTH DDWR A THAN'[1]

W elsh coal and its unique qualities for steam-raising purposes, particularly in the development of steam navigation, was a major factor in the development of docks and shipping facilities in South Wales. A supplier of the coal preferred by the early steamship pioneers was Llangennech Coal Co. who obtained an Act of Parliament to build a connecting railway and a floating or wet 'collier-dock', seen as the first of its kind.[2] Despite having considerable quay accommodation on the River Loughor, the dock, opened in July 1834, could accommodate ships of up to 1,000 tons and has been regarded as the first of the public floating docks in Wales.[3] This was the Llanelli New Dock, which was always '… kept full of water by means of lofty stop-gates, formed of African oak…' The works, consisting of dock, tide-basin and ship canal, were designed by George Bush (c.1810–41), later to be Brunel's resident engineer on the Taff Vale Railway (TVR).[4] Previously docks in and around South Wales were built as an improvement to facilities for canal development and for shipbuilding and shiprepairing. An early example of the former was the short canal, or tidal cut, built in the 1690s by Sir Humphrey Mackworth (1657–1727) from the River Neath to his works at Melincryddan. By 1699 a pill was constructed with strong flood-gates at its entrance in order to accommodate larger vessels, referred to as 'a docke' by Mackworth in his private diary on September 1696. [5] Apart from this pioneer work, it was not until 1798 with the Glamorganshire Canal's sea-lock that this innovation was seen again. As the first on the Severn Estuary, it allowed ships of up to 200 tons access to the mile-long floating basin through the 97ft by 27ft sea-lock near the mouth of the River Taff.[6] The sea-lock at Cardiff freed the port from the constraints of the tidal estuary, although not completely as the winding two-mile approach channel was completely dry for three hours a day.

Cardiff at that time was not dominated by the export of coal even though it had the greatest advantage over all the South Wales ports due to its position and the hinterland it served, namely the widest catchment of the central part of the coalfield. Cardiff had the advantage based on the convention that the natural resources of a valley should be exported in the direction of the flow to the sea.[7] By the 1830s coal was exceeding iron in tonnage shipped, a trend that would not be reversed at Cardiff, and coal was, as we have seen, the trigger for development elsewhere.[8] Other 'modern' floating docks, served by railway rather than canal or tramroad and authorised under an Act of Parliament, followed the Llanelli New Dock; there was Port Talbot, in 1837, Cardiff's Bute Dock in 1839 and Newport in 1842. As well as an increase in the instance of dock building, the size of docks and lock entrances would also increase, and dock owners were put under considerable pressure to keep up with the increasing size of ships, particularly paddle steamers coming into service. Brunel played a major role on both sides here, both in developing steam navigation and designing many innovative dock works. Dock work figured early in Brunel's engineering career; Monkwearmouth dated from November 1831, then there were surveys for the navy yard at Woolwich and, what would turn out to be beneficial in terms of outcomes for Brunel himself,

Above right: *Captain, later Admiral, William Henry Smyth (1788–1865). Smyth moved to Cardiff as Lord Bute's advisor on the construction of Bute Dock and was to lead Bute's opposition to the shipping place on the River Ely proposed by the TVR, publishing his* Nautical Observations on the Port and Maritime Vicinity of Cardiff *in 1840. (Smyth, W.H.,* The Sailor's Word-Book: The Classic Dictionary of Nautical Terms*).*

Above left: *Wrth Ddwr A Than, which translates as 'Through Water and Fire', was the motto of the Cardiff Railway Co., formerly the Bute Docks Co., successors to the trusteeship that had managed Bute Dock on behalf of the Bute family. It was formed to change the status of the Bute Docks Co. to that of a dock-owning railway company, but, despite the name, the docks, one of the great seaports of the world, remained the most important part. This can be seen on the side of the Pierhead Building at Cardiff Docks. (SKJ photograph)*

work for the Bristol Dock Co. This work was overshadowed when his railway commissions, in particular the Great Western Railway, took off. His 1870 biography includes a chapter on 'Dock and Pier Works', but the subject of all of his dock works does not rate a separate chapter, even in the engineering critique edited by Sir Alfred Pugsley in 1976, or indeed any other book since published.[9] Opening the 1870 chapter on dock and pier works, his son simply states in Victorian formality; 'Mr. Brunel's dock and pier works are interesting, …' this comment is based on the details of their construction, the use of wrought iron and also that his dock at Monkwearmouth was one of the earliest of his independent designs.[10] The dock and pier works attributed in this biography are Monkwearmouth, Bristol, Plymouth (Millbay), Briton Ferry, Brentford and Neyland.

Chronologically, the first work of this kind for Brunel in South Wales was for a dock at the mouth of the River Ely, for which coal-loading places took the form of timber jetties or 'staiths' (which Brunel described as 'staithes and shipping places'[11]) which would later be substituted.[12] The case against this was led by Captain W.H. Smyth, who condemned the siting of such a dock (referred to as a Cogan Pil), because of tidal currents and the difficulties when onshore easterly winds prevailed. It was, he claimed, '… a very bad loading place…', adding that '… no steam tug of the capacity at present used at Cardiff, would be able to tow a laden vessel from the Pil round the Point…'[13] Smyth would also criticise Brunel on the question of staiths, particularly in that benching in the river had not been considered in order to accommodate vessels having to take the ground when the tide was on the ebb, otherwise the slope of the river bed would cause ships to tilt over at an angle unsuitable for loading. If the ships were to be kept afloat they would have to be moved out into the deep water in the middle of the river – in which case they could not use the staiths without the running out of coaling drops into the water. Smyth believed this was a proceeding that '… Mr. Brunel never seems to have even contemplated – taking the ground and loading there is what was contemplated …'[14]

The estimate of the staiths was another point of dispute with Smyth as Brunel doubled the cost from £8,000 to £16,000. When questioned on the increase in his 1836 estimate, Smyth claimed that Brunel's answer was; '… I shall have more staiths.' Smyth points out that all the other ports, such as Seaham and Sunderland, were adopting docks instead of staiths, but that the great mineral district of Glamorgan was to be opened up; '… by a new port erected with these staiths, which every body else has rejected…'

Twenty-five years later a dock was built, by this time the problems of tide and wind overcome by steamships steaming in and out of the dock or sailing vessels, as in the case of a certain ship in 1886, assisted by the ready availability of steam tugs.[15] A general history of engineering written by the engineering historian J.P.M. Pannell (1899–1966) devotes a chapter to 'Docks

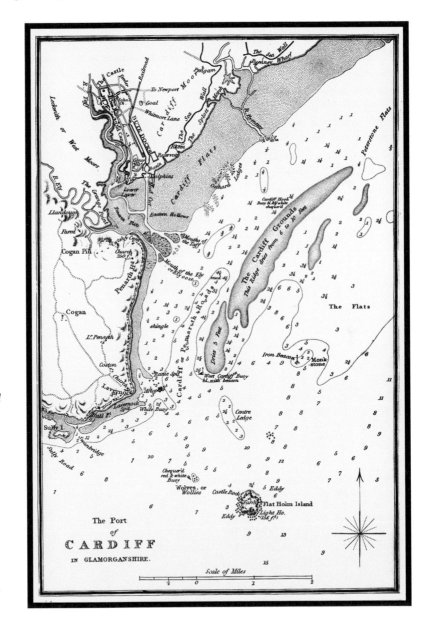

Map showing the approaches to Cardiff, published in Smyth's Nautical Observations on the Port and Maritime Vicinity of Cardiff *(1840). In this book he attempted to denounce the practicality of building a dock on the River Ely at Penarth.*

and Harbours'.[16] A member of the GWR's Civil Engineering Department, he was engaged on railway and dock engineering work from 1922 to 1936. Pannell comments on Brunel's docks at Briton Ferry, Millbay and Brentford, saying that they are not works of great magnitude, but that the wrought-iron gate designs by him are of interest '…and led to improvements in the design of semi-buoyant dock gates.' The number of projects involving dock, shipping and pier works by Brunel would run into double figures, but those actually completed would be somewhat less. His first Welsh dock proposal on the River Ely was for the TVR, and because of Brunel's involvement with that railway and the later South Wales Railway (SWR) at Cardiff, both of which had connections with the Bute Docks, it might be assumed that he had been involved with dock development here. This appears to have been the assumption made when a contemporary engineer wrote Brunel's obituary for the Institution of Civil Engineers (ICE). An obituary notice lists just three of his dock works:

> His introduction to Bristol led to his appointment as Engineer to the docks of that city, which he materially improved. He had been previously engaged in the construction of the old north dock at Sunderland, and subsequently, he was consulted about the design for the Bute Docks at Cardiff.[17]

The last reference to the Bute Docks was corrected in the following Minutes of the Proceedings of the ICE (Vol.20), as the first of the Bute Docks at Cardiff was originally planned by James Green (1781–1849), with revisions and consultations from Thomas Telford (1757–1834), then William, later Sir William, Cubitt (1785–1861) and Robert Stephenson. The dock opened in 1839 under the superintendence of one of Telford's former assistants, George Turnbull (1809–89), and the chief contractor was Daniel Storm who was also a contractor on the TVR.[18] The association of Stephenson and Bute dated from 1832, Bute engaging him in the early 1840s to advise and report on engineering matters connected with the dock and

Left: *Canton, or Cardiff, Bridge over the Taff. In 1850 Brunel turned the Taff downstream of the bridge to accommodate his SWR, which cut off access to the old port area of the town. However, in this engraving ship's masts can still be seen right of St John's church tower. (Mr and Mrs S.C. Hall,* The Book of South Wales, the Wye, and the Coast, *1861)*

Opposite: *Cardiff, showing the old bed of the River Taff following Brunel's diversion, St John's church tower in background. (Mr and Mrs S.C. Hall,* The Book of South Wales, the Wye, and the Coast, *1861)*

to negotiate and oversee the TVR connection to the west side of the Bute Dock. John Scott Russell also claimed to have advised Lord James Stuart, the brother of Lord Bute and who frequently acted on his brother's behalf in his affairs, around 1839 on a salt sea marsh on the banks of the Taff.[19] Bute sought Stephenson's input, presumably to match him against Brunel, even though the Bute Dock was his first dock engineering involvement as a consulting engineer.[20] Stephenson appointed Henry Swinburne (1821–55) as resident engineer at Cardiff, an engineer who later worked on the Chester & Holyhead Railway and surveyed many of Stephenson's foreign railways.[21] Brunel was to have several meetings with Stephenson over the Ely Dock proposal and the location of coaling appliances on the Bute Dock, but he had no input into the design of the Bute Docks, even though two of his railways served, firstly, the first Bute Dock (later known as the West Bute Dock) and then the East Bute Dock.

Running almost concurrently with his shipping proposals for the TVR was Brunel's interests in the development of an Irish packet port, specifically Porthdinllaen on the Llyn peninsular, intended to be the destination for Brunel's broad-gauge high-speed railway, the Worcester & Porthdinllaen, but destined never to be built.[22] Such activity links Brunel with Charles Blacker Vignoles at around the same time William George Owen was appointed to Brunel's staff. Owen had carried out a survey of Porthdinllaen for a proposed harbour at that time and in 1843 Owen returned under Brunel's direction to resume this earlier work which included soundings at Porthdinllaen for a packet pier and breakwater. Like the Worcester & Porthdinllaen Railway, these proposals for a major harbour at Porthdinllaen that would supersede Holyhead as the major Irish traffic port would come to nothing. Despite this setback, Russell and Brunel kept Dublin, the Irish objective of the Worcester & Porthdinllaen, in view as a SWR objective by supporting Irish routes, although as C.R.M. Talbot once exclaimed: 'He had never satisfactorily explained to him upon what grounds it could be supposed that through passengers would travel to Dublin round by Fishguard and Wexford.' Porthdinllaen has been described as '…a place which nature intended to have a history

CARDIFF.

but which never did'.[23] The same could be said for Abermawr in South Wales which was considered as a suitable western terminus and port by Brunel. Fishguard, or Goodwick as the present port is more properly known, was the first choice as one of the original two western termini of the SWR, but it lost out to Abermawr, with Fishguard being revived at the turn of the century by

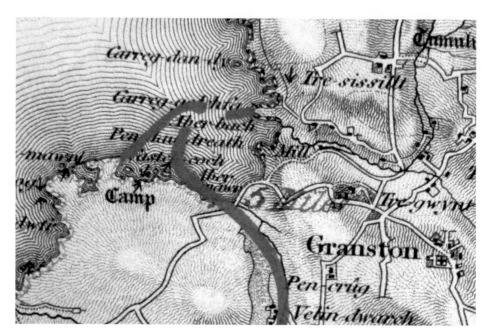

Map deposited by the SWR for the diversion of the line from Fishguard to Abermawr. The Parliamentary Bill of 1848, however, did not get the necessary support and was dropped but not before some work was undertaken. On this plan the two breakwater arms were drawn in red, enclosing the two coves of Aber Mawr and Aber Bach. The line from the SWR is also drawn in and shown as extending along a pier halfway along the rocky headland of Aber Mawr, jutting into the enclosed harbour and directly protected by the long breakwater arm 'A'. (Courtesy of the Pembrokeshire County Record Office)

Aberbach at Abermawr, early twentieth century. This photograph was taken from Carreg Golchfa, before the last major storm (winter 1922/23) which caused this part of the coastal road to fall into the sea and washed away the shore ends of the telegraph cable. (Courtesy of Roger Worsley and the Tregwynt estate)

the GWR. Abermawr itself was to lose out to the third and final choice of Neyland. This chain of events began in the autumn of 1847 when Captain Claxton was called upon by Brunel to investigate the merits of Fishguard. He surveyed the Irish Channel minutely to ascertain the best route to Ireland: '… and the elaborate survey then made appears to have led to the abandonment of Fishguard as the terminus, and the adoption of Abermawr, a few miles distant from Fishguard in a western direction.'[24]

The proposed harbour at Abermawr Bay, some four miles west of Fishguard, would have shortened the passage to Ireland by about five miles and the line connecting it by about a mile and a half. The harbour was to be made by enclosing the two coves of Aber Mawr and Aber Bach; a long breakwater (arm 'A') would extend from the rocky headland on the side of the Aber Mawr, known as 'Penmorfa'. On the opposite side was a much shorter breakwater (arm 'B') that was to extend from the rocky side of Aber Bach, known as 'Carreg Golchfa'. The railway was to extend along a pier halfway along the rocky headland of Aber Mawr that jutted into the enclosed harbour and was directly protected by arm 'A'. None of this work would be built, although the SWR actually began working on this alternative to Fishguard before parliamentary approval was granted, the Bill submitted to Parliament in 1848 did not get the necessary support and was dropped.[25] Today both Aber Mawr and Aber Bach are stranded bays with a shingle bank formed by the force of a great storm in 1859 known as the Great or Royal Charter Storm.[26] But that storm was to occur some twelve years after the decision to abandon Fishguard for Abermawr and then Abermawr itself and the branch to Pembroke Dock, the latter being abandoned in favour of Neyland on Milford Haven. Abermawr would have some connections to a wider world when the Irish link with the Atlantic telegraph cable came ashore here in 1862, but that's another story.[27] In the financial situation the SWR found themselves in in the late 1840s, Neyland was seen as a cost-effective solution that allowed the other western termini, Pembroke Dock, otherwise known as Pater or Paterchurch, to be abandoned or at least put in abeyance, and served by ferry from Neyland.[28] So the natural harbour of Milford Haven would still be reached even if it was at a different terminus from the one that had been proposed for the SWR in 1845.

Neyland, or, as it was known in this 1861 image, New Milford. In the background can be seen the station, railway activity and beyond that, the Neyland Hotel. (Mr and Mrs S.C. Hall, The Book of South Wales, the Wye, and the Coast, *1861)*

It was in the surveying of the Pembroke branch that Brunel realised that the line could be going through the property of his former mentor Nicholas Roch (1786–1866), who had retired to Paskeston Hall near Cosheston in Pembrokeshire.[29] It was Roch who had supported Brunel in securing positions at Bristol, including work for the Bristol Dock Co. In October 1844 he discovered that the line of the railway was on course to go through the house of his old friend,and he instructed his engineer, Robert Brodie, to '…try and keep the line a little further from Mr. Roches', informing Roch on 8 November 1844 that '…a singular chance or fatality has carried my levels almost thru' your house…' But this line, going through Roch's house or otherwise, would not be built by the SWR.[30] The destination of Pembroke Dock, however, was regarded as a 'token' gesture by some of those with an interest in the development of Milford Haven. One anonymous author whose 1846 pamphlet was entitled *Remarks on the Pre-eminent natural resources of Milford as the Western Terminal Port to the Railways of Great Britain* naturally promoted the merits of Milford Haven for such a port. In the contemporary argument then taking place about good and safe harbours on the western coastline of Great Britain, he pointed out that there was scarcely a prospectus of any projected railway in Wales, and there were plenty to choose from that did not include some form of harbour provision as part of its intended destination.[31] Why then did the SWR, the most important railway proposal of that time, not fully appreciate Milford Haven's superior facilities? The author levelled the following claim against the SWR:

> The engineer overlooked it, the directors knew it not, and having satisfied the claims of the Admiralty and the Post Office by a Branch to Cocheston [sic] Creek above Pembroke Ferry, the South Wales Line sweep round it to form a harbour according to the highest rules of art at Fishguard.[32]

Interestingly, when this pamphlet was revised and republished the following year (1847) the above comment was omitted. Perhaps there is a realisation that any railway connection with Milford Haven was to be encouraged, even a 'token' link to Pembroke Dock in light of the 'want of internal communications', and that a barrier that would soon be '…broken down by means of the South Wales and other projected, Rail Roads.'[33] Brunel again sought Claxton's advice when exploring the inlets and bays of the Milford Haven waterway. With Claxton's help his choice was Neyland, and, with the terminus finally determined, Brunel sought to provide dock accommodation there. An Act of 1852 provided for the development of the west bank of Westfield or Neyland Pill. Three years later the SWR submitted a Bill to develop the east, or Barnlake, side of the Pill, to allow for the enclosure of the Pill to create a floating dock. In this proposal passenger traffic would be kept to the west side while a connecting goods railway fifty-two chains long would run down the east side. Estimated at £18,000, the Bill for the proposed scheme included a clause to '… appropriate such purposes as they think fit for the waters of Neyland Pill which would otherwise flow into the Milford Haven.'[34] C.R.M. Talbot (1803–90), the SWR chairman, expressed confidence when these proposals were raised, as in 1853 he had been inundated with plans from interested parties using Milford Haven:

> Plans had been proposed for constructing docks both on the south and north side at an expense of about 500,000 and he had little doubt that ultimately both projects would become apparent for extending the railway to those docks, and thus making it the highway to America. (Hear, hear.)[35]

However, there were objections. The landowner Lord Milford, Richard Bulkeley Phillips/Grant (1801–57), reconsidered his willingness to sell the required land if the Pill was dammed, and he would only go ahead if a lifting bridge was substituted for the dam to allow boats to enter the

Bullo Pill Dock taken on 5 October 1946 showing the basin partly silted up at almost high-tide. The tip, although still standing, is derelict, having last been used in the early 1930s. Originally a drawbridge spanned the dock entrance allowing access to more tips alongside the river at Box Wharf. (Ian Pope collection)

Entrance to Bullo Pill Dock in 1978, abandoned and in a state of dereliction. The dock has recently undergone refurbishment and is being used for ship repairs, etc. (SKJ photograph)

creek at Westfield Pill. This could probably have been negotiated but a greater force, and one that had wrecked Brunel's plans on previous occasions, joined the opposition. The Admiralty, in support of the Royal Navy dockyard at Pembroke Dock (which had refitted Brunel's *Great Western* steamship fifteen years earlier), claimed that the damming of the Pill would result in the silting up of Milford Haven at the point used by the Royal Navy for turning warships using the dockyard. They also objected to the building of jetties associated with the development that would impede naval movements. Without the dock, development interest in Neyland dwindled away, and Brunel therefore had to abandon his proposal and concentrate on developing the west bank of the Pill. With this scale of setback and the lack of any meaningful volume of trade using Neyland, cross-Channel or transatlantic, Talbot was forced to admit in 1856 that '…the vision of greatness' he had entertained just three years earlier had faded away.[36]

At the other end of the SWR, on the Forest of Dean branch, there was work to be done at a dock – hardly the result of great ambition but rather a modest dock on the Severn at Bullo Pill. The dock here came about following the appointment of John Rennie (1761–1821) by the Department of Woods and Forests to advise on transport development as it affected their interests. In 1807 he described 'Bulloe Pill' as capacious and conveniently approached when he proposed that a 'little wet dock' should be built there.[37] Partly opened by June 1810, it consisted of a modest stone-lined basin approximately 90 yards long and 20 yards wide for which Brunel would design coaling appliances and a swingbridge across the lock gate entrance when the branch (formerly the Bullo Pill Railway) was converted from a 4ft-gauge tramroad to broad gauge in 1854.[38]

The opening ceremony for the new South Dock at Swansea was performed by Charlotte Emily Talbot, Talbot's eldest daughter, just over a week after Brunel's death on 23 September 1859. The Swansea Harbour Railway, for which Brunel had prepared plans based on the ideas put forward in his 1846 report for a high-level line to facilitate access to the dock, also opened at the same time. William Tredwell (1819–71), from a family of civil engineering contractors, was responsible for the railway contract which cost £70,000. He also completed the South Dock after the undertaking passed to the Neath Harbour Trustees.[39] William's father, William Tredwell Sr, and his four brothers all undertook engineering contracts for Brunel; his youngest brother Soloman (1822–59) was to construct the launching ways for the *Great Eastern*.[40] Construction of the high-level railway began in August 1857 and the quays along the north dock were '… altogether transformed by the erection of the substantial arches; the canal wharves, the Brewery premises, Richardson's dry dock …' The last named was to almost wreck the transfer of the dock to the trustees as the Swansea Corporation carried a clause in the enabling Bill then before Parliament that the dry dock should not be taken without written consent of the corporation. From Brunel's engineering viewpoint, all or part of the site occupied by the dry dock had to be taken in order to construct a workable railway, '…and therefore, the South Wales Railway would not abide by their arrangement with the Trust if the clause were retained.'[41] An urgent meeting led to both councillors and ratepayers joining in a written declaration that the clause would be withdrawn; to the relief of the promoters the Bill, minus the clause, became law. Brunel could now proceed with the high-level railway continuing from the SWR's Swansea Coal Branch (thirty-four chains long) that had opened in June 1852 from a junction just outside the SWR High Street station. There would be much work in Swansea for the hydraulic machinery made by William George Armstrong, later 1st Baron Armstrong of Cragside (1810–1900).

Armstrong had shown great interest in mechanics and engineering as a child, but was articled to a solicitor, although he began investigating hydraulic and hydrostatic power while practising law.[42] The first orders for his new hydraulic machinery included hydraulic warehouse lifts, and later two cranes for the Liverpool Docks. James Meadows Rendel (1799–1856) was an enthusiastic supporter of Armstrong's venture, whose apprentices would include Henry Marc

Brunel (1842–1903), and recommended the adoption of hydraulically operated dock gates and cranes at Grimsby Great Dock in 1850.[43] The applications hydraulic power could be used for were extended further, but one order for hydraulic cranes set a problem for Armstrong as he could not use local water mains pressure or construct a tall water tower,[44] a problem resolved by the invention of the hydraulic accumulator, in which a heavy weight was raised by a pumping engine which, when raised to the full height, acted as a ram forcing the water in a cylinder down, thus giving the required water pressure. Water pressures could be increased from a few hundred pounds per square inch (psi) to 600 psi and pressures of 1,000 psi were being used by 1858. Brunel became an enthusiastic supporter of hydraulic power used for the coal containerisation system developed to minimise coal breakages which was derived from coal boxes or tubs used by Thomas Powell (1784–1864) on the TVR. On 2 September 1851 Brunel sketched a number of ideas for the iron boxes which in matured form consisted of 4ft 8in cubes loaded four at a time onto special platform trucks at the pit and hauled to the docks. An insight into the hectic working life of Armstrong can be seen from this letter to his wife in October 1850:

> At York at 7 p.m. on Sunday night, at New Holland at 11 next morning, at Grimsby at 3, and in London at 11 at night. Today at Rendel's but found he was at Birkinshaw. At Fowler's who wants more machinery at New Holland – at Brunel's who I expect will require hydraulic coal drops – at Pickford's where I hope for cranes and at the present time I am preparing for a meeting at the Engineers. This is the multum in parvo of my proceedings…[45]

The Armstrong works was not confined to the manufacture of hydraulic equipment as it undertook general engineering work of all descriptions and would grow to become one of the most successful engineering and armaments businesses in Victorian Britain, and indeed the world. From the time of the Crimean War, Armstrong was developing new types of artillery such as breech-loading rifled guns, although these would be manufactured by a separate

Swansea South Dock in 1865, some six weeks after the opening by Charlotte Emily Talbot. The first combination drop and wagon tip was working on 5 November 1859. The special loading towers for the container wagons can be seen in this photograph. See fuller view in Vol.II, p.215. (Courtesy of the Royal Institution of South Wales)

company.[46] He was also looking at the application of wire-wound barrels, which both he and Brunel had been investigating since 1855. Indeed, Brunel was to give Armstrong a commission to make such a gun, but an existing patent precluded any further work until the patent expired.[47] Henry Marc Brunel's letter books contain numerous references to the Armstrong gun and the rifle trials between Armstrong and Whitworth. HMS *Warrior*, built in 1860 by Scott Russell, had 110-pounder Armstrong guns, making it not just Britain's first iron-clad battleship but the largest and fastest warship in the world, capable of destroying any other vessel afloat. Armstrong asked Brunel on 9 March 1852 if the engine house originally designed for Swansea would be applicable at Neath; 'as we cannot arrange the pipes &c until this is determined.' Armstrong concluded his letter: 'We shall very soon be ready to despatch the materials for Swansea & Neath. They will be sent by sea.'[48] In October 1852 Armstrong visited Swansea to check one of his drops that Brunel thought was not working properly:

> I am happy to say that a few hours practise with the Drop at Swansea after you left it on Wednesday removed all difficulties & the man was enabled to work it smoothly, quickly & without any bungling. The two boxes brought up on each wagon were repeatedly shipped in 3 minutes being 1½ minutes each box including the hooking & unhooking.[49]

There were further complaints in August 1853 regarding the breakage of two or three jibs at Swansea. Armstrong believed that only one had failed '...due to a nut not having been applied properly.'[50] In his 1846 report to the Duke of Beaufort, Brunel condemned the project of converting the new cut into a floating dock, arguing that a new dock should be built on the western side of the harbour (which would be taken up as the South Dock). However, the Vale of Neath Railway (VNR) would have been grateful for the drops on the North Dock as these and access to a wharf on the River Neath, opened by the Briton Ferry Docks & Railway Co. from the SWR in April 1852, were the only shipping facilities available to the VNR in its first year of opening. Hand-operated cranes were also used, particularly at the smaller quays and shipping places, and a 3-ton crane was supplied to Carmarthen Quay in about 1850 by Stothert of Bath, a company that had supplied broad-gauge locomotives to the GWR. Stothert's also supplied the engines for pumping water from the hollow cast-iron cylinders of the Wye Bridge at Chepstow in order to sink them under pressure, and, as Stothert & Pitt, the company supplied the first steam crane to be used at Swansea in 1865.[51] At Swansea, the VNR would supplement the drops at the North Dock themselves but, despite this the accommodation open to them, they remained limited and the Swansea option could only be accessed over SWR metals. To get to Swansea, VNR coal trains had to be pulled up, and usually split because of the gradient, a tortuous 1 in 90 incline at Skewen, by a mixed bag of GWR engines, GWR providing the motive power and rolling stock to SWR.[52] Some six weeks after the opening of the South Dock on 5 November 1859, a combination drop and wagon tip was available there which could handle boxes and tipped wagons, and in 1862 the VNR were to take a long lease of the Swansea Harbour Railway after the SWR had withdrawn from a similar arrangement. A dock at Briton Ferry, or Baglan Bay as it is first referred to, is seen as a solution for the VNR and a proposal for a floating dock there was raised in 1850.[53]

Before any public reporting of the 'New Floating Dock and Tidal Cut at Baglan Bay, Neath Harbour', Joshua Williams, the secretary of the VNR, estimated the cost at £44,233, and by the end of May a number of backers, including Talbot, J.H. Vivian and Hussey Vivian, indicated their support.[54] They are not present at a meeting held at the VNR offices on 22 June 1850, but the following promoters of the proposed docks were: Griffith Llewellyn (of Baglan Hall), John Rowland (the Neath banker), Evan Evans (1794–1871), Henry Simmons

Briton Ferry from Warren Hill. The dock entrance is not visible in this Victorian promotional image that was used by Robinsons of Bristol for printing on paper bags. (Owen Eardley collection)

Enlargement of lithograph of Briton Ferry Dock in 1861. See colour section for full image. (Courtsey of the Neath Antiquarian Society)

Coke, and Joshua Williams.[55] The first two were local investors and partners in projects such as County Banks, whilst Coke, the solicitor, attorney and town clerk of Neath, was instrumental in proposing the railway that became the VNR for which Joshua Williams would become secretary in 1849.[56] He also had outside industrial interests, but Evan Evans was Briton Ferry's most entrepreneurial son, involved in a wide range of ventures, from breweries to railway contracting and coalmining.[57] The meeting discussed the proposal on behalf of the landowner, George Childs Villiers, 5th Earl of Jersey (1773–1859), with the suggestions of his agent, Richard Hall (1806–78), as to the terms for the docks and a modification of such terms was requested, subject to communication with the SWR as to the plans and operations being prepared by Brunel. Hall had worked on the Bristol & Gloucester and later the South Devon Railway, becoming responsible for negotiating the purchase of railway land, and was proud of his friendship with Brunel. Another man who had worked under Brunel and who would become adopted son of Briton Ferry as a prime mover in not only building the dock but also in taking up the largest number of shares was William Ritson (1811–92).[58] Coming from a family of contractors from the north-east of England, Ritson first worked for Brunel in 1838 and work on the VNR included the Pencaedrain Tunnel contract near Hirwaun and then the troublesome Merthyr. A company was formed for the building and operation of the dock, with a capital of £60,000 divided into 3,000 shares of £20 each.

Ritson would take some time off work at Hirwaun to write to Joshua Williams on 23 August 1850, enclosing his signed subscription form for £10,000 of shares (he was the largest private shareholder).[59] While Ritson was considering his investment, the committee met again, and on 31 July 1850 they noted that the terms for a lease of the ground necessary had been submitted and that a copy of the amended and altered clauses were to be sent to Hall who they were prepared to meet regarding '…the subject at Chepstow on Saturday next.' Brunel's plans and report were also discussed at the meeting, with the option of two plans. Plan No.2 was seen as the preferred option. However, the extra works necessary for this option, in '…forming the six upper shipping places and the outer gates of the half tide Basin with such parts of the masonry necessary thereto…' and '…without interrupting the Trade, so as to confine as far as possible the first outlay within Mr. Brunel's estimate for No. 1' was seen as prohibitive.[60] Plan No.2, however, required additional land and the day after Brunel reported on the cost of such a dock (on 9 August 1850, supported by the SWR and VNR) he met with Lord Jersey to discuss the land required and sketch the apportionment of the land, a strip some 150ft wide between the boundary of the company's land and Warren Hill. This land, amounting to half an acre, was in the possession of Henry Scale of Briton Ferry Ironworks who agreed to give it up and place it in the hands of Lord Jersey, '…in aid of the dock scheme.'[61] The sketch was left with Jersey who then wrote a note on it, presumably to send on to Richard Hall:

> Mr. Brunel has just left me. He supposes that there may be some discussion with Mr. Vivian about the excavation of the Warren Hill – if that be the case He thinks we should be satisfied with the space of 150 ft. as marked by Him in pencil above. He thinks we should press about time for decision. He believes Mr. Vivian & Mr. Talbot eager to have the Docks & thinks the additional £15,000 for plan No. 2 will be raised.[62]

A full report was produced by 27 September 1850 with notice being served regarding an application to Parliament for the dock and railway.[63] On 2 November 1850, following a further meeting of the Docks Committee, it was noted that having consulted with Brunel and Lord Jersey, the advice of their solicitor was not to proceed under the Joint Stock Companies Act but to apply for an Act of Parliament in the next session.[64] This was arrived at after consider-

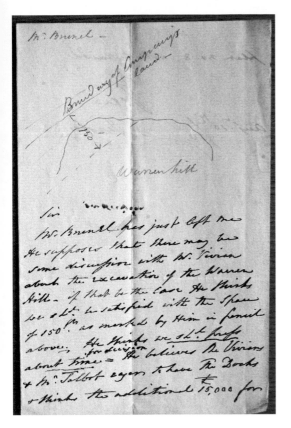

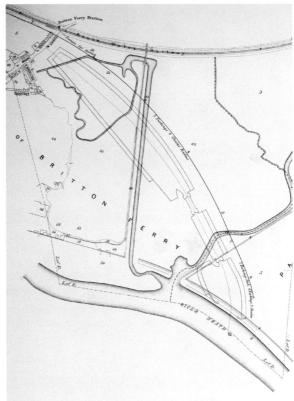

Above left: *A meeting with Brunel and Lord Jersey on 9 August 1850 concerns the acquisition of additional land for the No.2 option at Briton Ferry. Brunel makes a sketch which is left with Jersey who then writes a note on it, presumably for his agent, Richard Hall, beginning, 'Mr Brunel has just left me...' (Courtesy of West Glamorgan Archive Service)*

Right: *Design No.2 by Brunel for Briton Ferry. It offered extra shipping places, an entrance basin and entrance channel with three sets of lock gates, allowing greater access to the dock; however, the additional expense could not be met and Design No. 1 was proceeded with. (Courtesy of the Neath Antiquarian Society)*

ing the implications of proceeding without an Act for seven months, assuming the earliest that such an Act could be obtained would not be before July (and then it noted 'with influence'), expending upwards of £10,000. They also resolved that a deputation wait upon the VNR to solicit them to become subscribers to the project; '...for any such amount, not less than £10,000, as they may deem advisable.' The 'minimum' sum of £10,000 was based on the guarantee of £500 a year in dock tolls that the VNR had proposed; the biggest investor was to be the SWR at £20,000, with Ritson at £10,000, the Earl of Jersey at £5,000 and other private subscriptions at £15,000, making up a capital of £60,000. John Rowland was the chairman of the committee on this occasion and he signed off the final resolution of the minutes with which they would communicate their intentions with the railway companies '... to Mr. Brunel...' The minutes expressed their desire to make an early start on the scheme as:

> The committee feel that on the Vale of Neath Line a large demand for Steam Coal must of necessity spring up in the district and that the Ports which can afford accommodation to the

Briton Ferry Dock with the lock gate closed. (Sandra Davies collection)

shippers will at once make a trade, legitimately belonging to Briton Ferry, from its contiguity, provided accommodation could be given at that place. From this fact combined with other circumstances, this committee do consider it as highly desirable the works of this undertaking should be commenced as quickly as arrangements can be made for doing so without deferring the same until the act is obtained provided the Railway Company would not act conjointly with the Promoters in doing so.

By April 1851 a Bill was in the House of Commons, reaching the Lords a month later, and in July 1851 it became law as 'An Act for making and maintaining Docks at Baglan Bay … with a branch Line of Railway to the South Wales Railway.'[65] A railway that would bring coal traffic to this dock was also discussed in the August of the following year, 1852, as the Maesteg, Glyncorrwg & Briton Ferry Railway which, by November of the same year, was turned into the South Wales Mineral Railway (SWMR) and notice was given of its intention to go to Parliament.[66] Before work began at Briton Ferry, stages for shipping by basket were ordered to be erected in May 1852, but these proved to be inconvenient.[67] The first reference to engineers and surveyors being engaged on the site of the proposed docks was in May 1853, with Ritson commencing work in August of that year.[68] In March 1854 work on the dock was reported as 'advancing steadily', with Brunel and Joshua Williams inspecting the works in September of the same year. [69] Joshua Williams reported to the board of the VNR in March 1855 that he frequently suggested to Brunel that simple tipping apparatus to employ ordinary wagons should be used '…in conjunction with the cranes at Briton Ferry…' The board agreed and Armstrong was asked to supply tipping machinery to be erected at Briton Ferry.[70] Back at the dock site several problems delayed the progress of construction, such as in 1855 when a severe labour shortage affected South Wales. This was a situation covered in the national press; *The Times* referred to the large amount of railway and dock work being undertaken in South Wales and the inadequate supply of labour. Because of this, the rates of wages were then running some 25 to 30 per cent above ordinary prices. 'Common labourers', it stated, were obtaining 3s 6d a day, and skilled men like blacksmiths could get up to 6s a day, whilst shipwrights could get even higher wages. Highlighting the situation, it reports that: 'The wages of the men employed in the construction of the Briton Ferry Docks have been increased to 3s. 4d. per day, and this solely in consequence of the difficulty experienced by the contractors in securing a sufficient number of hands.'[71]

Briton Ferry Dock; an aerial photograph taken in 1934. The Albion Steelworks is in the foreground and beyond it is the Baglan Bay Tinplate Works which became the machinery store for the Albion in 1956. On the other side of the dock is the Briton Ferry Steelworks. (Sandra Davies collection)

The tugboat Goodwill of Bristol *operated on the River Neath and at the dock for many years. In this photograph from the 1920s coal wagons are still waiting to be unloaded, but these are the twilight years of the dock. (SKJ collection)*

The navigation from the sea to the dock also involved Brunel, who submitted a report in March 1856 to the Neath Harbour commissioners meeting at the Town Hall.[72] This involved improving the navigable channel of the River Neath at its embouchure into the Bristol Channel. It had been said that the course of the river in the 1820s ran in a zig-zag form with vessels seen, when going out, making a tack to north, east, west and south. Henry Robinson Palmer had originally formed a bank of furnace slag to direct the course of the river as far seawards as it was felt practicable. This, referred to as the 'slags', the point marking the Neath Bar, was proposed as the site of a new lighthouse in order to improve navigation into the River Neath in *The Cambrian* report of 17 May 1850. This bank helped to improve the treacherous bar at the river mouth, and between 1842 and 1843 Alfred Russell Wallace (1823–1913) was engaged in taking soundings of the river between the town and the sea.[73] Brunel carried a training bank still further, and succeeded in cutting off a bend of the river, this work being continued by his former chief assistant, Robert Pearson Brereton (*c.*1818–94), who extended the navigable channel in a straight line to low-water mark, a distance of two miles, and lowered the bar to within a foot of the level of the dock sill.[74] In 1898 it was stated that since

Far right: *Lock gate recess with the partially dismantled lock gate, taken in 1978. (SKJ photograph)*

Right: *Briton Ferry, a RAF survey photograph of 1943 which shows the entrance piers at the mouth of the river. (Courtesy of Welsh Assembly Government Aerial Photographs Section)*

Briton Ferry, a RAF survey photograph of 1952. The A48 bridge can be seen in the course of construction across the end of the dock. (Courtesy of Welsh Assembly Government Aerial Photographs Section)

In this photograph taken in 1978 from the end of the dock where the A48 bridge crosses overhead, the Albion Steelworks can be seen. This would close in 1980. (SKJ photograph).

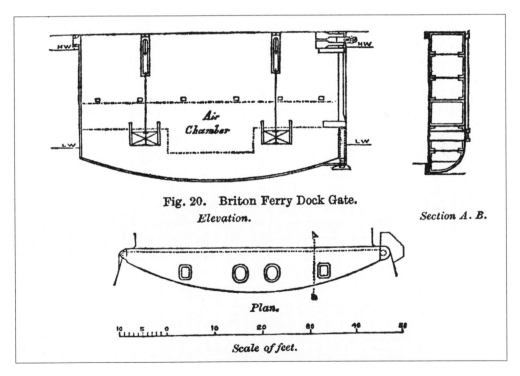

The wrought-iron buoyant gate at Briton Ferry, decribed in The Life of Isambard Kingdom Brunel (1870) as 56ft long, 31ft 6in deep in the middle and 26ft 6in at the sides. It was constructed by Edward Finch of Chepstow, the company who, as Finch &Willey of Liverpool, had come down to Chepstow to build Brunel's Wye Bridge.

then the services of a dredger were never required in the river or its entrance to the channel, the course being kept clear by its own natural tendency.[75] On 18 April 1856 Ritson was advertising in *The Cambrian* for tenders to supply Aberthaw cement to be delivered to his wharf on the river, and later in the same year he sacked Irish navvies reported as being at the centre of a row during the building of the dock. Apart from minor incidents like this, Ritson's financial position was somewhat exposed and on 30 August 1856 a scheme was put forward in an effort to make payments owed to him through shares and debentures.

Henry Austin Bruce (1815–95) chaired a meeting of the directors on 22 September 1856, referring to his endeavours to get the unallotted shares taken up, calling on his father John Bruce Pryce to take fifty, with H.H. Vivian, J.H. Rowland, G. Llewellyn and Richard Hall also agreeing to fifty each. George Thomas Clark (Bruce's fellow Dowlais ironworks trustee) agreed to take ten, and the long-suffering Ritson was persuaded to take 100.[76] Ritson, no doubt, believed it was only by continuing to invest in the scheme that he would get what was owed to him for the contract, let alone a return on his shares. That left 360 shares, which Hall announced would be taken by Lord Jersey in the name of Lord Villiers. Shortage of funding caused the VNR to support a Bill to empower them to raise money in May 1857, and in August of the same year £15,000 was advanced by the VNR.[77] On 27 July 1857 the new Act was passed empowering them to raise money and for other purposes.[78] Calls were made on shareholders and a new share issue was announced on 12 March 1858, but there was a further slip at the dock works in July 1858, adding 'extra cost to construction' and Brunel reported on the work necessary to complete the dock in August of that year.[79] A year later another call was made on shareholders and in the same month *The Cambrian* (19 August 1859) reported on the extent of the financial problems bedevilling the completion of the dock. On 10 August 1859 Brunel's clerk Bennett wrote to Williams enquiring, on Brunel's behalf, '... the financial position of the Dock Company. I need hardly remind you of the very large amount that is owing to Mr Brunel by which he is much inconvenienced.[80] ' The following month, which marked the death of Brunel, also saw a final call made on shareholders, but on 14 October 1859 Ritson was forced to auction off some of his plant and materials.[81] December of that year saw further criticism of both the Briton Ferry Dock and the VNR, the former not having been completed and the latter being badly run, and a report followed in January 1860.[82] Work began again and later in July 1860 a lease of the dock was authorised with regard to the VNR, and new shares were offered to the proprietors of the VNR in August 1860.[83] Things were still low, however, and the September 1860 half-yearly meeting had to be adjourned due to low attendance. But on 2 November 1860 it was announced that the docks were about to be completed, and in the same month the SWMR proposed a Bill for extending their line to the docks, and, as Brunel's replacement as the SWMR engineer, Brereton was referred to as 'Mr Brunel's assistant'. [84] Finally it was announced that: 'the spacious dock will be opened on August 23 1861' – a date brought forward by one day to 22 August.[85] Brunel's professional charges were not settled before his death or even before the dock was opened and a bond (to the sum of £7,075 3*s* 6*d*) had to be taken out between the dock company and Brunel's widow, Mary Elizabeth Brunel, and his son Isambard. Similar arrangements had to be undertaken after Ritson's death to his executors.[86]

The dock, when completed, consisted of an inner basin and an outer tidal basin with a single lock gate, along the lines of Brunel's plan 1, rather than plan 2 that proposed a smaller tidal harbour with a lock entrance to a short inner dock leading to a double lock channel to the larger dock. The dock consisted of an outer tidal basin with a total area of 10 acres, and an inner floating basin of 14 acres.[87] A coffer dam was not used in the construction of the work but a large bank of slag and earth was used as a sea dam and was cut through when the works were completed, a channel dredged to a depth of 6ft below low water.[88]

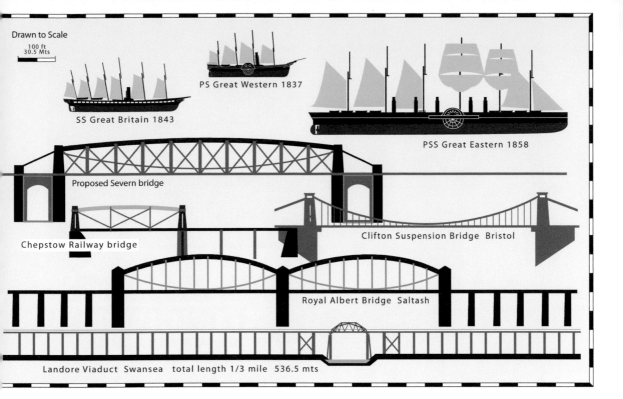

Drawn to Scale

100 ft
30.5 Mts

PS Great Western 1837

SS Great Britain 1843

PSS Great Eastern 1858

Proposed Severn bridge

Chepstow Railway bridge

Clifton Suspension Bridge Bristol

Royal Albert Bridge Saltash

Landore Viaduct Swansea total length 1/3 mile 536.5 mts

1. Above: *Building in iron or timber, above water or on it, Brunel pushed the boundaries of engineering technology. On land he built his longest timber viaduct at Landore (a third of a mile long from end to end) and shortly afterwards his widest (in 1852) iron span at Chepstow, the design of which would lead to his largest iron bridge with two 455ft spans at Saltash in 1859. Not completed in his lifetime was his first major commission, the Clifton Suspension Bridge, and never attempted was his proposed bridge across the River Severn with a 1,100ft span. On sea his entry into steam navigation began with the wooden paddle steamer* Great Western *continuing with the revolutionary iron-hulled and screw-propelled* Great Britain *to the leviathan that was the* Great Eastern. *(Graphic design by Owen Eardley)*

2. Left: *A close-up of Brunel as one of the participants in John Lucas's painting of the 'Conference of Engineers at the Menai Straits to Floating one of the Tubes of the Britannia Bridge'. (Below) 6 Robert Stephenson, 7 Mr. Hemingway, 8 Captain Claxton, 11 Joseph Locke and 12 Brunel. See Vol.II, p47, for a view of the complete painting. (Courtesy of the Institution of Civil Engineers)*

Isambard Kingdom Brunel

Charles Dickens

SUN ALLIANCE & LONDON
INSURANCE GROUP
Bartholomew Lane, London EC2N 2AB

People have been bringing their insurance problems to us for more than 250 years. Famous people, and not-so-famous people, with all kinds of problems . . . of life and limb, home and property, business, industry and travel. And they keep on coming to us today–

3. Born in the same town and christened at the same font, Brunel and Charles Dickens also took out insurance with the Sun Alliance and London group! Part of a 1976 advertising campaign. (SKJ collection)

4. Close-up of Captain Christopher Claxton (1790–1808) with flat hat and glasses, again from John Lucas's painting. Claxton met Brunel in the early 1830s and they become lifelong personal friends and colleagues. He was to become the managing director of the Great Western Steamship Company, and Brunel would rely upon him in all matters maritime, including the floating out of iron railway tubes and trusses, assisting Robert Stephenson in this role at the Menai Straits. Standing next to him is the contractor John Hemingway (b.1802) who was engaged on the construction of the piers for the Britannia Tubular Bridge. (Courtesy of the Institution of Civil Engineers)

5. Right: *The stained-glass window was installed in 1954 to illuminate the main lobby of the Institution of Civil Engineers' building in London. On this is inscribed Henry Robinson Palmer's role in the formation of the institution, with armorial bearings above. (Courtesy of the Institution of Civil Engineers)*

6. Below right: *Map of Porthdinllaen with the intended harbour and destination of the Worcester & Porthdinllaen Railway, drawn with soundings and sketches of the breakwater position by William George Owen (1810–85). This was published at Caernarvon in 1836 just before he joined Brunel's staff. (Courtesy of the Institution of Civil Engineers)*

7. Below: *The South Wales Institute of Engineers crest on the original President's Board at Bay Chambers, Cardiff. The Welsh inscription translates as 'I gave to man the wings of an eagle', a saying attributed to Edward Williams, better known as the poet and antiquary Iolo Morgannwg (1747–1826). His grandson, also Edward Williams (1826–86), was one of the founders of the institute and occupied the post of secretary from its inception until 1864. Up to that year, Williams worked under William Menelaus (1818–82) at Dowlais, where he claimed the honour of rolling the first Bessemer steel rail. Williams returned to become president of the institute from 1881 to '83. (Courtesy of the South Wales Institute of Engineers Educational Trust)*

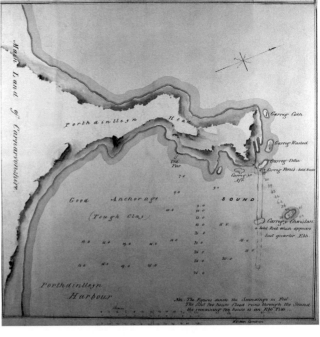

8. Above: Portskewett, the site of the Bristol & South Wales Union Railway (B&SWUR) pier on 25 April 2007. In this photograph, looking towards the second Severn Crossing, the railway once emerged from a cutting on to the timber pier extending 708ft into the Severn. Of the pier only a few timber stumps remain (one protruding section can be seen at the edge of the water) and even the GWR boundary marker has gone. (SKJ photograph)

9. Right: Plaque celebrating the opening of the B&SWUR in 1863. Originally erected on the portal of the down-line tunnel at Patchway, this plaque has gone from being inaccessible and almost impossible to view to being on show to the hundreds of railway passengers that pass through Bristol Temple Meads each day. Christopher Thomas (1807–94) was born in Llangadog in Carmarthenshire and, as a wholesale grocer, butter merchant and haberdasher, carried out much trade with Bristol and eventually moved there to set up a soap manufactory. He became chairman of the company in 1863, and the first purpose-built steamer operating the ferry was named after him. (SKJ photograph)

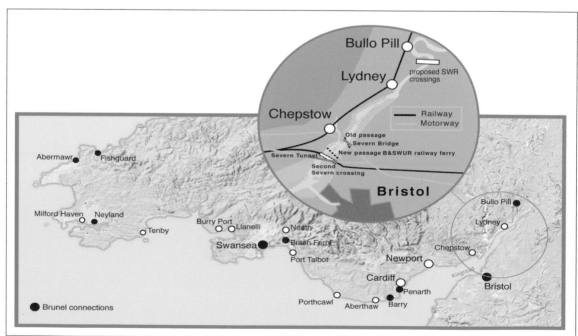

10. Left: The masonry abutment of the New Passage still remains on the Bristol side and part of the hotel can be seen in the background. The hotel had been acquired by the B&SWUR company as part of the land and property it needed, and was sold on to a separate company who would provide facilities for passengers waiting for the ferry boats. On days of wild and windy weather this could be longer than expected! Working here and at Portskewett, the resident engineer Charles Richardson first began thinking of building a tunnel under the Severn. (SKJ photograph)

11. Right: Plaque to Charles Richardson in St Mary's church, Almondsbury, taken on 2 July 2008. (SKJ photograph)

12. Below, right: Plaque on the masonry abutment looking out at low tide, the pier being 1,635ft long here. Prior to closure, the daily timetable showed eight passenger trains running each way and two each way on Sundays, between the pier and Bristol Temple Meads. Both views of New Passage taken on 18 May 2007. (SKJ photograph)

13. Below: The Severn Railway Bridge in May 1967. It graphically illustrates the damage done on that fateful night in October 1960. (SKJ collection)

14. Opposite below: Ports and crossing points along the Severn and South Wales coast. In terms of shipping places and ferry crossings, Brunel was associated with Abermawr, Briton Ferry, Bullo Pill, Cardiff (railway links), Fishguard, New Passage, Neyland, Pembroke Dock, Penarth, Portskewett, Porthcawl and Swansea. The brown dots show his major involvement, although Fishguard, Abermawr and Pembroke Dock would be abandoned in favour of Neyland, and his son, Henry Marc, would work at Penarth and be responsible, with John Wolfe Barry, for Barry Docks. The inset of the Severn shows the various crossings at the points between the New and Old Passages and Brunel's rejected SWR viaducts. Despite his proposals for timber viaducts and even a tunnel, and later an iron tubular suspension bridge, only his steam railway ferry (part of the Bristol & South Wales Union Railway) would be built, to be opened after his death. (Cartography by Owen Eardley)

15. South Wales Railway £20 share certificate, issued to John Parry de Winton of Maesderwen, Brecon, on 30 June 1855, and signed by the company secretary Frederick George Saunders (1820–1901), who had taken over as secretary when Captain Nenon Armstrong absconded in 1849. (SKJ collection)

16. Dock development at Swansea and Briton Ferry. The Swansea map was actually revised in November 1926 and appeared in the following year's edition of Great Western Ports, whilst the aerial photograph of Briton Ferry was taken by the RAF and combined from two negatives in 1949. Superimposed on this is the line of the M4 viaduct, which resulted in the infilling of the dock at this point, and the A48 viaduct which skims the dock, although part of this extreme end had previously been filled in. (Cartography by Owen Eardley)

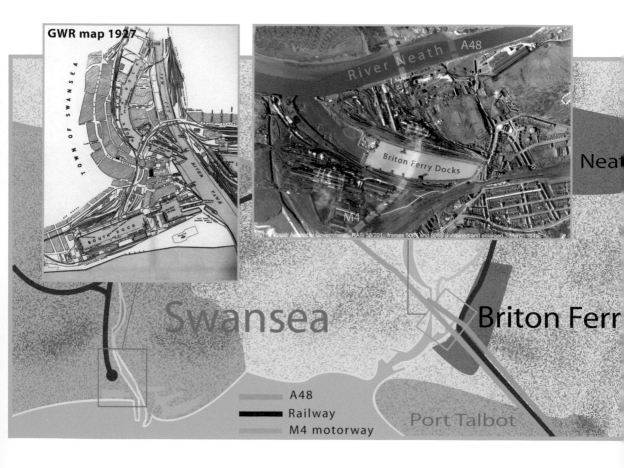

17. Above: *Coloured lithograph of Briton Ferry Dock depicted as opened in 1861. A SWR passenger train can be seen in the foreground with the hydraulic tower and coaling appliances alongside the dock. The curious six-sided building in front of the tower is listed simply as 'stores' on later plans. Most of the ships are sailing vessels, with a steam-paddler towing one ship through the lock entrance with what looks like a steam dredger at the entrance to the dock. The vessel with a single funnel proceeding up the River Neath is believed to be* Neath Abbey, *a pioneer iron-screw steamship built by the Neath Abbey Iron Co. in 1846. An enlargement of this can be found on page 89. (Courtesy of the Neath Antiquarian Society)*

18. Below: *Modern view of the dock on 28 October 2008. The now restored hydraulic tower can be clearly seen alongside the dock comprising of a silted-up and overgrown inner harbour and outer harbour at low tide. In the distance Mumbles can be seen, and at the entrance to the dock the remains of the timber mole or breakwater structure, and opposite, in somewhat better condition, a masonry mole. The lock entrance is now filled in and in shadow in this photograph, but a section of Brunel's lock gate remains. (SKJ photograph)*

19. Above: *Chain journey: the chain cable for Brunel's* Great Eastern *steamship was made here at Ynysyngharad, otherwise known as the Newbridge Chainworks of Brown Lenox at Pontypridd. The tested chain cable was loaded into canal boats in the works' canal pond which then proceeded down the Glamorganshire Canal to Cardiff, a sea journey taking them around the south coast and up the Thames Estuary to Millwall. From the bottom left-hand corner of the photograph the canal runs diagonally, and beyond the bushes, the disused Ynysyngharad Locks can be clearly seen. In this April 1977 view the testing shop and many of the original chainsmiths' shops can be seen. (SKJ photograph)*

20. Below: *The* Great Eastern *on the stocks. A stern view of the ship under construction at Millwall, from a coloured lithograph in the* Pictorial History of the Great Eastern Steam-ship, *published by WH Smith in 1859, before her first Atlantic crossing. (Courtesy of Margaret Greenwood)*

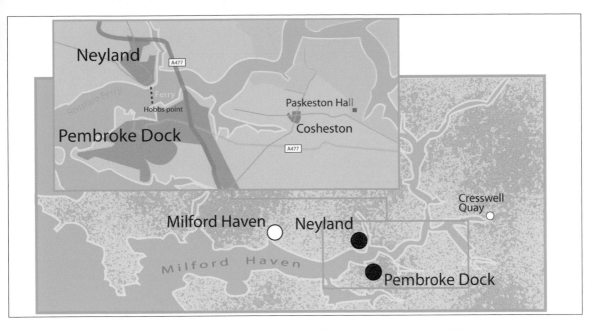

21. Above: *Milford Haven showing Pembroke Dock and Neyland, Neyland becoming the terminus of the SWR and the home port of the Great Eastern. The inset shows the location of Paskeston Hall in Cosheston, the home of Brunel's friend and mentor Nicholas Roch (1786-1866) which narrowly avoided Brunel's proposed line to Pembroke Dock! (Cartography by Owen Eardley)*

22. Below: *Paskeston Hall, Cosheston. In a letter to Roch in November 1844 Brunel admits that '…a singular chance or fatality has carried my levels almost thru' your house…' (University of Bristol, Special Collections PLB Brunel to Roch, 8 November 1844). On 28 October 1844 he had instructed his assistant, Robert Brodie, to keep the line '…a little further from Mr. Roches'. The line was abandoned in favour of a single terminus at Neyland, and although the SWR attempted to keep the powers alive, another railway company via a different route would complete the branch to Pembroke Dock. (SKJ photograph courtesy of David and Lorraine Scheeres)*

23. Above: *PSS Great Eastern underway in an oil painting by J. Studdard, 1857. (Courtesy of David A. Cross)*

24. Below: *Neyland, a hand-coloured photograph from a stereocard pair taken when the Great Eastern was on the gridiron. (SKJ collection)*

25. Above: *Site of the gridiron at Neyland on 20 May 2008. One of the distinctive buildings opposite in Pembroke Dock can just be made out in the background of this 1860 photograph showing the* Great Eastern *on the gridiron (see photograph on page 144). The building on the left is the Neyland Yacht Club, and a row of terraced houses called 'Great Eastern Terrace' faces the site. (SKJ photograph)*

26. Below: *Neyland, looking towards the Milford Haven Bridge on 14 September 2004. The bridge superseded the ferry from here to Hobbs Point, Pembroke Dock, in the 1970s, and the floating pier for the car ferry that replaced Brunel's was demolished. Barlow rail can be seen as fencing on the right and timber work that made up the protective 'dolphins' can be seen out of water on the left. (SKJ photograph)*

27. Above: *Railway lines serving the former station and quay area at Neyland can still be seen in 2004. The original broad-gauge lines continuing to Brunel's floating pier of which his plan, signed and dated 10 July 1855, can be seen in the inset. Plan courtesy of the National Record Office, Kew. (SKJ photograph)*

28. Below: *Milford Haven Dock on 6 December 2008. In 1875 the* Great Eastern *returned from cable-laying duties to be laid up herself here at Milford Haven. Above the docks on the horizon can be seen the Lord Nelson Hotel on Hamilton Terrace. Brunel stayed here when planning the facilities at Neyland. On Hamilton Terrace a stone, now also marked with a plaque, measures the length of the ship as the distance from St Katherine's church. The docks were built around the* Great Eastern, *and it was a tight squeeze when it came to getting her out. (SKJ photograph)*

29. Above: *The SS Great Britain leaving Penarth Dock on her last voyage, 6 February 1886: a reconstruction by Owen Eardley of the scene in 1886 to mark the centenary of her last voyage in 1986. The ship is being towed out of the dock by a paddle tug, and the Docks Office and Customs House can be seen on the left. (SKJ collection)*

30. Below: *The SS Great Britain back home in Bristol on 20 June 2009, restored in the dry dock where she was built and launched in 1843. Moored nearby is the replica of the sailing ship;* Matthew. *(SKJ photograph)*

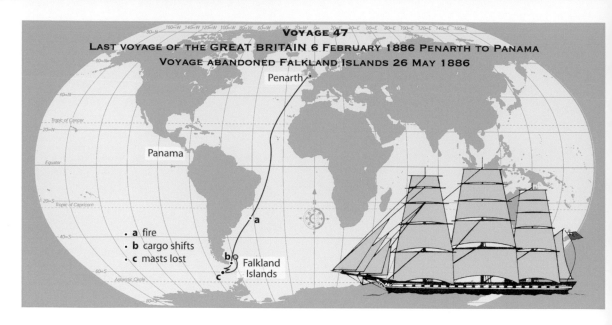

VOYAGE 47
LAST VOYAGE OF THE GREAT BRITAIN 6 FEBRUARY 1886 PENARTH TO PANAMA
VOYAGE ABANDONED FALKLAND ISLANDS 26 MAY 1886

Penarth

Panama

a fire
b cargo shifts
c masts lost

Falkland
Islands

31. Above: *Voyage 47, the last commercial voyage of the* Great Britain. *(Cartography by Owen Eardley)*

The return Journey from the Falklands Islands to Bristol.
On 25 March 1970 the salvage team arrive in Port Stanley and on 7 April the Great Britain *is refloated and four days later berthed on to the salvage pontoon* Mulus III. *On 24 April the ship leaves the Falklands on the pontoon towed by the German tug,* Varius II *and arrives in Montevideo, Uruguay, on 2 May. The world's media gathers to record the return of the* Great Britain *across the Atlantic on 6 May. On 22 June she reaches Barry roads in the Bristol Channel. The ship arrives at Avonmouth on 23 June and is floated off the pontoon on 1 July to be towed up the River Avon on 5 July and returns to her Dry Dock on 19 July. The two week wait in the Cumberland Basin, was in order for a high enough tide to get her back through the locks into Bristol's Floating Harbour and back to the Great Western Dockyard where she had been built.*

32. Below: *A watercolour of Penarth Dock by G.H. Andrews, 1865. This gives a detailed view of the dock and the river staithes in the River Ely, and also a good impression of the landscape familiar to Henry Marc Brunel when he worked on the dock under Sir John Hawkshaw. An enlargement of the lock entrance can be seen on page 162. (Courtesy of Glamorgan Record Office)*

33. Left: *Hydraulic accumulator at Tower Bridge, taken on 10 February 2008. Another commisssion undertaken by the John Wolfe Barry and Henry Marc Brunel partnership, although their work, particularly the constructional engineering of the bridge, is encased by the architectural masonry of Horace Jones (1819–87). Henry Marc came into his own with regard to the hydraulic equipment that operates the bascule spans and which can be viewed today as part of the visitor experience of Tower Bridge. Hydraulic power was provided by water, at a pressure of 750psi, from several hydraulic accumulators. The system was supplied and installed by the company where Henry Marc had served his pupilage, Armstrong of Newcastle. Each accumulator used a 20in ram driven down by a very heavy weight to maintain the desired pressure. A similar system would have been housed in the accumulator tower at Briton Ferry. (SKJ photograph)*

34. Below: *Barry Docks on 10 July 2009; a view from Clive Road on Barry Island. Working with John Wolfe Barry, this is the most significant engineering works, in terms of scale, by Henry Marc Brunel. Looking across the first stretch of water is the Commercial Dry Dock, 867ft long and 100ft wide, then the Lady Windsor Entrance Lock, a deep lock of 647ft long with a 65ft-wide lock entrance. This allows vesels to enter the docks at all states of tide. The inner gate of the three lock gates is also visible in the photograph. Alongside this is No.3 Basin, a tidal entrance into the main docks. All vessels enter the No.1 Dock, comprising 62 acres and opened in 1889, with access to No.2 Dock through the Junction Cut to the right. This 34-acre dock was opened in 1898 and a ship can be seen alongside the Rank Flour Mill in the distance. The masonry and red brick building on the left, looking back across No.1 Dock, is the Barry Dock Office, with a statue of David Davies (1818-90) in front. The inscription refers to him as 'Deputy Chairman and Chief Promoter of the Barry Dock and Railway Company'. Apart from the masonry abutments, none of the coaling appliances remain. See photographs on page 180 showing shipping activity here in the first quarter of the twentieth century. (SKJ photograph)*

35. Left: *The statue of Brunel at Neyland. Sculptured by the late John Thomas of Barry, this statue shows Brunel with a broad-gauge engine in one hand and the 'leviathan' in the other. All, of course, fitting for the location. This work was the last executed by Rhondda-born Robert Thomas, an artist who has sought to remind former industrial communities of their rich heritage. His works include Nye Bevan at Cardiff and the miner and his family, a group that commemorates the scene of the 1910 Tonypandy riots. It was unveiled after his death in 1999 on the redeveloped site by Prince Charles. Sadly, in the early hours of Monday 23 August 2010, the 8ft high bronze statue was stolen from its plinth at Brunel Quay. It is hoped that the stature will be replaced in the near future using the original sculptors moulds. (SKJ photograph)*

36 and 37. Below: *Close-ups of the 'leviathan' and the broad-gauge engine. (SKJ photograph)*

Entrance to the dock from the River Neath showing the M4 and A48 bridges. Photograph taken on 20 May 2009. (SKJ photograph)

The GWR station at Briton Ferry. (John Alsop collection)

The two basins were connected by a passage or entrance of 50ft in width with entry to the basin through a single lock gate, a Brunel–designed wrought-iron buoyant design, 56ft long, consisting of a buoyant gate with five vertical bulkheads and six decks.[89] Finch & Heath of Chepstow were the gate contractors and Edward Finch appears to have used the experience of constructing these gates when setting up an engineering business in Cardiff.[90] Brunel had observed that the lock gates he designed for Monkwearmouth, constructed of yellow pine timber with a great bulk of light wood at the bottom of the gates, gave them a certain amount

The Whitford Point lighthouse, an unusual cast-iron design dating from 1865. Built under the Llanelly Harbour Act of 1864, it had gone out of use in 1921, but when fitted with a solar-powered light in the 1980s it became the only operational wave-washed lighthouse in the British Isles. It was supplied by the contractor who had carried out much work for Brunel on the SWR, George Hennet. (SKJ photograph)

of floatation. This buoyancy, at the lower part of the gate, was '…somewhat analogous to that of the air-chamber which Mr. Brunel introduced afterwards in his wrought-iron gates.'[91] His first wrought-iron buoyant gate was designed for Bristol Docks in 1847, and could be moved at ease by chains attached to the barrels of powerful crabs. The first ship scheduled to enter Briton Ferry dock was a barque which was 'unfortunately delayed at Swansea', so the pioneer iron screw steamship *Neath Abbey* took the honour of being the first vessel into the new dock.[92] After the official opening ceremony, 200 guests were entertained to a celebration breakfast presided over by the Neath Harbour commissioner, and the Mayor of Neath, Alexander Cuthbertson. The VNR and the SWMR brought coal from the valleys for shipment, building up a thriving trade in both coastal and export shipments. The *May Stetson* was the first vessel to be loaded with coal at Briton Ferry, with its cargo of Aberdare coal destined for the *Great Eastern* at Liverpool, and at the celebratory breakfast the late Mr Brunel was praised, along with Lady Jersey and her agent, Richard Hall.[93] Ritson, who acted as the dock contractor under very difficult financial circumstances, has left a number of reminders of his work in Briton Ferry. Ritson Street commemorates him, and there at the entrance to that street stood the Wesleyan chapel, designed and constructed by him, as was the old Briton Ferry National School standing on the site where a Roman Catholic church now stands.[94]

Equipment installed at the new dock consisted of one hydraulic coal-tip with a lift of 40ft, three cranes with a lifting capacity of 1½ tons each and one ballast-crane capable of moving 20-30 tons per hour, and in 1870, nine years after it opened, some 2,398 vessels had used the dock.[95] This was not enough for it to be profitable and the dock company went into receivership in 1872.[96] In 1873 the dock was taken over by the GWR, following which trade steadily declined, with the GWR favouring the use and development of Port Talbot and Swansea Docks. There were also long drawn-out delays with another railway, the Rhondda & Swansea Bay Railway, in reaching the dock, which it did not achieve until 1893.[97] Trading at the dock continued into the 1930s, but after this it was used only by vessels of the South Wales Sand & Gravel Co. and as an anchorage for the naval training ship used by the local sea cadets. The dock was officially closed in 1951 but continued to be used by sand boats until 1959 when the last ship, the *Glenfoam*, left after discharging its cargo at the South Wales Sand & Gravel wharf.[98] Part of the lock gate remains in situ and the hydraulic accumulator tower, in which the Armstrong accumulator was installed, was recently refurbished on 15 June 2007.[99] As an engineer, Pannell was proud to be able to claim a link through his father, who trained under James, later Sir James, Inglis, to Brunel. Inglis, as the GWR's chief engineer, finally linked the GWR to a new harbour at Fishguard in 1906, the original main terminus of the SWR having been abandoned over fifty years before. But, to give the last word on the dock to Pannell: 'That of Briton Ferry is out of use, the dock now being a tidal basin and little used. The author had the privilege of working on the gate when it was last repaired in the year 1929.'[100]

CHAPTER 5 NOTES:

1 Wrth Ddwr a Than, 'Through Water and Fire', the motto of the Cardiff Railway, formerly the Bute Docks Co., successors to the trusteeship that had managed the Bute Docks on behalf of the Bute family.

2 Lewis, Samuel (1840), *A Topographical Dictionary of Wales Vol.I* (S. Lewis: London). Llanelly entry.

3 Baber, Colin and Williams, L.J., Ed. (1986), *Modern South Wales: Essays in Economic History* (University of Wales Press: Cardiff). Section A; Symons, M.V., 'Coal-Mining in the Llanelli area – Years of Growth', 1800–64, p.61.

4 Lewis, Samuel (1840). Llanelly entry, see Vols I and II for George Bush.

5 Jenkins, Elias, Ed. (1974), p.239 and p.241, *Neath and District; A Symposium*, Chapter 12, Communications, Robert Thomas (Elias Jenkins: Neath). The 'docke' reference implies it was then under construction – Phillips, D. Rhys (1925), p.327, The History of the Vale of Neath (D. Rhys Phillips: Neath).

6 Rowson, Stephen and Wright, Ian. L. (2004), p.236, *The Glamorganshire and Aberdare Canals, Vol. 2, Pontypridd to Cardiff* (Black Dwarf Publications: Lydney). It opened on 1 July 1798 with the *Castle of Cardiff*, a sloop of 80 tons burden, arriving from Bristol.

7 Brunel ignored this when his Vale of Neath Railway (VNR) tapped into the top of the Cynon Valley and exported its mineral resources not down but up and over the top of the valley and then down the Vale of Neath. See Vol.II, Chapter 8.

8 In 1830, regarded as a depressed one for the iron industry, 87,372 tons of iron and 113,753 tons of coal were shipped. Lewis, Samuel (1840), Cardiff entry.

9 Brunel, Isambard (1870, reprinted 1971), *The Life of Isambard Kingdom Brunel* (Longmans, Green & Co.: London, 1870, reprinted by David & Charles: Newton Abbot, 1971). Pugsley, Sir Alfred, Ed. (1976), *The Works of Isambard Kingdom Brunel* (Institution of Civil Engineers and University of Bristol: London and Bristol). The situation was not changed with the wealth of new titles in 2006 (Brunel 200).

10 Brunel, Isambard (1870, reprinted 1971), p.417.

11 See Vol.I, p.119. For a description of the development of staiths see Powell, Terry (2000), *Staith to Conveyor: An Illustrated History of Coal Shipping Machinery* (Chilton Ironworks: Houghton-le-Spring).

12 See Vol.I, Chapter 9. This dock, however, or the erection of staiths on the Ely for the TVR, would not be built by Brunel.

13 Smyth, W.H. (1840), pp.13–14, *Nautical Observations on the Port and Maritime Vicinity of Cardiff* (W. Bird: Cardiff).

14 Smyth, W.H. (1840), pp.95–96.

15 The dock was built to the design of another engineer, albeit assisted by a young Henry Marc Brunel. Cardiff did not possess steam tugs until the opening of the Bute Dock. Smyth, W.H. (1840), p.74.

16 Pannell, J.P.M. (1964), *An Illustrated History of Civil Engineering*, republished in 1977 as *Man the Builder:*

An Illustrated History of Civil Engineering (Thames & Hudson Ltd: London).

17 Obituary notice; Isambard Kingdom Brunel (Vice-President), 1806–59, Minutes of the Proceedings, Vol.19, Session 1859–60, pp.169–173. 1860.

18 Turnbull began his career under Telford on preparatory works for St Katharine Docks, London. This dock being one of the first to employ steam pumps in order to maintain the water level of the dock, capable of filling the 180ft- (55m) long entrance lock in just five and a half minutes. *Civil Engineering*, May 2007, Vol.160, Special Issue One; Thomas Telford: 250 Years of Inspiration, paper 15033; pp.48–55, St Katherine Docks, London – Telford's high-speed harbour by Mike Chrimes.

19 Cardiff and Merthyr Guardian, 18 November 1864, '…visited Cardiff 25 years earlier…' The site was later claimed to be the site where the Bute shipyard was built.

20 Bailey, Michael R., Ed. (2003), p.356, *Robert Stephenson – The Eminent Engineer* (Ashgate Publishing: Aldershot).

21 See Vol.I, p.170. Henry was the brother of the poet Algernon George Swinburne (1837–1909).

22 See Vol.II,m Chapter 2.

23 Talbot's remark was reported in *The Times*, 31 August 1849, Morris, Tom, Porthdinllaen; from Harbour Co. to National Trust, Morfa Nefyn, Gwasg Carreg Gwalch.

24 Lewis, Samuel (1849), p.378, *A Topographical Dictionary of Wales Vol.I* (Samuel Lewis: London). This is the account relating to the SWR (under Glamorganshire) and a similar reference can be found under Pembrokeshire in Vol.II of the same work, pp.295–296.

25 Davies, Desmond N. (1997), p.10, *The End of the Line: A History of Neyland* (Pembrokeshire County Council: Haverfordwest).

26 As well as the loss of over 400 ships, the storm also claimed the Royal Charter.

27 A storm washed away the road above Abermawr beach and also the shore ends of the two cables in 1922 or '23. The station was then abandoned and the telegraph hut and cottage returned to the Tregwynt Estate.

28 See Vol.II, Chapter 7 for further details on the abandonment of Fishguard in favour of Neyland, and pp.238–9 regarding the rival standard-gauge line promoted to Pembroke by the Pembroke & Tenby Railway (P&TR).

29 Roch was a prominent Bristol businessman from an old Pembrokeshire family who had played an important role in starting Brunel's career in that city. See Vols I and II.

30 University of Bristol, Special Collections, PLB, 3, pp.36, 171–172, and 197, 10 May, 28 October and 8 November 1844. The story was part of *Brunel in Bristol* by R.A. Buchanan (Essays in Bristol and Gloucestershire History, the centenary volume of the Bristol and Gloucestershire Archaeology Society, Ed. Patrick McGrath and John Cannon, 1976).

31 No author cited (1846), p.4, *Remarks on the Pre-eminent natural resources of Milford as the Western Terminal Port to the Railways of Great Britain* (C. Roworth & Son: London).

32 No author cited (1846), p.7.

33 No author cited (1847), p.8, *Remarks on the Pre-eminent natural resources of Milford as the Western Terminal Port to the Railways of Great Britain* (C. Roworth & Son: London).

34 Davies, Desmond N. (1997), p.15, *The End of the Line: A History of Neyland* (Pembrokeshire County Council: Haverfordwest).

35 *The Times*, 26 February 1853.

36 McKay, K.D. (1989), p.93, *A Vision of Greatness: The Story of Milford Haven 1790–1990* (Bruce Harvatt Associates, in association with Gulf Oil Great Britain Ltd: Haverfordwest).

37 Paar, H.W. (1965, second edition 1971), pp.37–40, *The Great Western Railway in Dean; A History of the Railways of the Forest of Dean: Part Two* (David & Charles: Newton Abbot).

38 See Vol.II, p.84.

39 Cross-Rudkin, P.S.M., Chrimes, M.M., and others Ed. (2008), pp.785–788, *Biographical Dictionary of*

Civil Engineers in Great Britain and Ireland, Volume II; 1830–90, Thomas Telford Publishing on behalf of the Institution of Civil Engineers: London. The cost of the dock, including the earlier phase of work undertaken by the contractor Joseph Pickering, was £169,073.

40 See Chapter 7.

41 Jones, W.H. (1922), pp.200–1 and 281. A plan prepared at the time, showing the dry dock, can be see in Vol.II, p.211.

42 McKenzie, Peter (1983), *W.G. Armstrong: A Biography* (Longhirst Press: Newcastle upon Tyne). In 1846 he was elected a Fellow of the Royal Society, sponsored by Faraday and Wheatstone.

43 McKenzie, Peter (1983), p.119. Armstrong took on six apprentices in the first year.

44 Grimsby Docks required a 300ft-high tower in order to produce the water pressure required.

45 McKenzie, Peter (1983), p.49, 17 October 1850.

46 This was the Elswick Ordnance Works, set up because of Armstrong's Government positions, on Government committees and appointments for the War Department.

47 The patent had been taken out by Longridge, who never implemented the idea.

48 University of Bristol, Special Collections, copies of Armstrong letters, DM 1306/x/4, letter from Armstrong to Brunel 9 March 1852.

49 Copies of Armstrong letters, DM 1306/x/8, Armstrong to Brunel 1 October 1852.

50 Copies of Armstrong letters, DM 1306/x/18, Armstrong to Brunel 30 August 1853.

51 Torrens, Hugh (1978), pp.50–52 and 58, *The Evolution of a Family Firm: Stothert and Pitt of Bath* (Stothert and Pitt Ltd: Bath). This Carmarthen crane, complete with iron jib, survived until 1953.

52 Eventually in April 1861 it was agreed to allow the VNR to operate its own coal trains over the SWR and finally for the VNR to operate its own line into Swansea; the Swansea & Neath Railway.

53 *The Cambrian*, 10 May 1850. Such a dock, as a Brunel/SWR project, had been proposed in 1846.

54 D/D BF/E 560, April 1850, West Glamorgan County Record Office. *The Cambrian*, 31 May 1850.

55 D/D BF/E 556, 22 June 1850.

56 See Vol.II, Chapter 8 for further details. He was to take over the Traffic and Locomotive Departments and the Engineering Department in place of Brunel in 1858.

57 Jenkins, Elias, Ed. (1974), p.208.

58 Cross-Rudkin, P.S.M., Chrimes, M.M., and others, Ed. (2008), pp.665–667.

59 D/D BF/E 575, 23 August 1850. Ritson was to take twice as many shares as the Earl of Jersey.

60 D/D BF/E 557, 31 July 1850.

61 D/D BF/E 575, correspondence file, Henry Jones to Richard Hall 2 November 1850.

62 D/D BF/E 575, Earl of Jersey 10 August 1850, some words, i.e., 'eager', may be incorrectly interpreted.

63 *The Cambrian*, 9 August 1850 and 15 November 1850.

64 D/D BF/E 558, 2 November 1850.

65 *The Cambrian*, 11 April, 16 May and 11 July 1851. Act ref; xlix (L. & P.) Baglan Bay (Briton Ferry) in Jones, T.I. Jeffreys, Ed. (1966), p.185, *Acts of Parliament Concerning Wales 1714–1901* (University of Wales Press: Cardiff).

66 *The Cambrian*, 13 August, 10 September and 12 November 1852. See Vol.II, Chapter 10.

67 Powell, Terry (2000), p.90.

68 *The Cambrian*, 13 May and 5 August 1853.

69 *The Cambrian*, 10 February 1854 reports the half-yearly meeting for taking place in Cardiff. The visits of Brunel and Williams are reported in *The Cambrian* for 29 September 1854.

70 The cranes were on the point of being finished in March 1854; Copies of Armstrong letters, DM 1306/x/26, Armstrong to Brunel 1 October 1852. Powell, Terry (2000), p.90.

71 *The Times*, 6 October 1855. What response follows an advertisement for a wheelwright in November 1855 is unknown (*The Cambrian*, 2 November 1855).

72 *The Cambrian*, 21 March 1856.

73 Phillips, D. Rhys (1925), pp.462 and 632. This was for docks on the River Neath proposed by Palmer but which were not finally started until 1874 and completed in 1886. See Vol.II, Chapter 8, for Wallace's role in the construction of the VNR.

74 Brunel, Isambard (1870, reprinted 1971), p.438.

75 Humphreys, E. (1898, reprinted 1982 by West Glamorgan County Council), pp.48–49, *Reminiscences of Briton Ferry and Baglan, E.* (Humphreys: Swansea).

76 D/D BF/E 559, 22 September 1856. Bruce would take the chairmanship of the VNR in 1859.

77 *The Cambrian*, 29 May and 21 August 1857.

78 Act ref; lxxix (L. & P.) 'Briton Ferry' in Jones, T.I. Jeffreys, Ed. (1966), p.186.

79 *The Cambrian*, 2 July and 20 August 1858.

80 West Glamorgan Archive Service, D/D PRO/BRB297, 29 November 1862.

81 *The Cambrian*, 2 September 1859.

82 *The Cambrian*, 16 December 1859. Meanwhile the improvements to Neath Harbour have been suspended since Brunel's death, *The Cambrian*, 23 March 1860.

83 *The Cambrian*, 20 July 1860, masons are sought to complete the works – to apply to J. Ritson at Glynneath or Mr Coulson at Briton Ferry. *The Cambrian*, 27 July and 24 August 1860.

84 *The Cambrian*, 16 November 1860.

85 *The Cambrian*, 19 July 1861 and 9 August 1861.

86 West Glamorgan Archive Service, D/D PRO/BRB 299-303, 25 July 1865.

87 GWR map of 1882 in Harries, Colin and Taylor, Felicity (2007), *An Archaeological Investigation at the Brunel Accumulator Tower, Briton Ferry, Neath Port Talbot* (Site Archaeological Services: Monmouth).

88 Brunel, Isambard (1870, reprinted 1971), pp.437–440.

89 Pannell, J.P.M. (1964, republished 1977), p.155.

90 *The Cambrian*, 23 August 1861. See also Chapter 4.

81 Brunel, Isambard (1870, reprinted 1971), pp.421–422.

92 Built by Neath Abbey ironworks in 1846, no doubt influenced by Brunel's *Great Britain*, and owned by three local men which included Evan Evans. Farr, Grahame, p.136 (1967), West Country Passenger Services, T. (Stephenson & Sons Ltd: Prescot).

93 Morgan, Cliff (1977), p.63, Briton Ferry (Llansawel) (Cliff Morgan: Briton Ferry). Decimus Prothero was the merchant for the coal, sourced from the Fforchaman colliery; *The Cambrian*, 23 August 1861.

94 Morgan, Cliff (1977), p.85. The chapel building later became the Steelworks Welfare Hall, but is now in use as a shop.

95 Morgan, Cliff (1977), pp.63–65. When first built, the dock extended as far as Dock Street itself and ships' bowsprits were known to reach over the roadway, even penetrating the window of one house on the occasion of a very high tide.

96 The GWR acquired the dock under its Act of that year and the Briton Ferry Act of 20 July 1873.

97 MacDermot, E.T., revised by Clinker, C.R. (1964), p.336, *History of the Great Western Railway, Vol.II, 1863–1921* (Ian Allen: London).

98 Morgan, Cliff (1979), p.81, *A Pictorial Record of Briton Ferry (Llansawel)* (Cliff Morgan: Briton Ferry). In 1969 it was bought by Neath Borough Council who intended filling the dock in and redeveloping it as an industrial estate, but this was put on hold at local government reorganisation in 1974.

99 The accumulator was completed and ready for installation sometime between December 1858 and September 1859, Harries, Colin and Taylor, Felicity (2007), p.4.

100 Pannell, J.P.M. (1977), p.155. Out of use, but not the end of the story, with the restoration of the tower being the first stage of a long campaign by Hugh James and others of restoring this unique dock work.

6

A MAMMOTH UNDERTAKING

'...THE IDEA OF A MAN FAMOUS FOR LARGE IDEAS

– MR BRUNEL.'[1]

Following their commercial success and with just two voyages behind it, the Great Western Steam Ship Co. considered building two ships of similar dimensions to the *Great Western* but eventually decided on a single sister ship. This ship would be another wooden vessel, and a shipment of African oak was purchased for the purpose, the company announcing the intended name of the new vessel as the *City of New York* in the late summer of 1838.[2] From this time on, however, proposals for the new ship began to change dramatically as Brunel was making calculations on the merits of iron over timber for the construction of the hull. Captain Claxton, as managing director of the Great Western Steam Ship Co., carried out experiments for Brunel with William Patterson on the iron coasting steamer *Rainbow*. They were suitably impressed with the results and reported in favour of constructing the new ship in iron. The adoption of iron and the increase in the size of the ship led to a new name, that of *Mammoth*. Work on the new vessel was announced in August 1839: '...a ship of iron, of about 2,000 tons tonnage, with engines of 1,000 horse-power.'[3] This, however, was not the end as far as changes were concerned; Brunel would bring in an even greater change to the plans when he decided to abandon paddle wheels and adopt the screw propeller. In 1840, Francis Petit Smith was promoting his patent screw propeller by practical demonstration on the *Archimedes*, a small vessel of 237 tons. Having failed to impress the Admiralty when it was demonstrated on the Thames, it was decided to send it around the coast in May of that year. Captain Chappell, RN, was to accompany the ship in order to thoroughly test the potential of the screw propeller, and on 29 May the *Archimedes* arrived in Bristol. On 1 June 1840 a number of prominent Bristolians were taken on a trip around the islands of Flat Holm and Steep Holm before returning to Portishead. Thomas Richard Guppy was onboard to examine the merits of screw propulsion when it subsequently left Bristol for Tenby, Milford and Liverpool, and he reported favourably.

The construction of the hull had already begun when this radical change was taken, and to adapt this newly invented form of propulsion to a ship, Brunel designed new engines and machinery, constructed by the company itself under Guppy's superintendence. The final form of the ship gave it a displacement of 3,618 tons and an overall length of 322ft, longer than any battleship then afloat, and it was launched with her third and final name at Bristol on 19 July 1843. Interestingly the Newport-registered *Great Britain* had been lost at sea on 30 March 1843 in a mid-Atlantic storm.[4] However, Brunel's ship had been referred to as the *Great Britain* for some time before this and there was little to compare the two ships;[5] Brunel's *Great Britain* being the first screw-propelled ocean-going iron ship, the forerunner of the modern Atlantic liner, whilst the Newport ship, to quote *The Bristol Mercury*, was of 404 tons burden, frigate-rigged and built at Quebec in 1839. The latter was also the first vessel to have entered the

MR. FRANCIS PETTIT SMITH,
FIRST PRACTICAL INTRODUCER OF THE SCREW-PROPELLER.—FROM
A PHOTOGRAPH BY LAWRENCE.—(SEE MEMOIR, PAGE 442.)

Francis Petit Smith brought the Archimedes *to Bristol in 1840 to promote the advantages of his patent screw propeller. Although Brunel was interested, it was Guppy who took a trip in the vessel to Tenby, Milford and Liverpool, and he was favourably impressed. (SKJ collection)*

Newport Town Dock, a dock that would not have restricted the passage of Brunel's ship.[6] But the *Great Britain*'s size would be a problem at Bristol soon after she was launched, or rather floated, out of the dry dock that she was built in. To gain access to the open sea Brunel supervised the widening of the lock entrance by an army of workmen so that the ship could be got out on the night tide of 11 December 1844, the last of the spring tides.[7] The first transatlantic voyage of 26 July 1845 began under the command of Captain James Hosken (1798–1885), and after some teething troubles the ship settled down on regular runs between Liverpool and New York. Disaster, however, struck the ship on 22 September 1846 when the *Great Britain* left Liverpool for New York with 180 passengers on her fifth voyage and her largest complement to date.

An almighty shudder run through the ship when, a few hours later and at night time, she ran aground in Dundrum Bay off the Irish coast.[8] It was a navigational error, for which Captain Hosken blamed his chart, but an error that would have severe consequences for the Great Western Steam Ship Co. There were no fatalities and the ship was not wrecked but it was to take some salvaging. Claxton rushed to the scene and a wooden breakwater was built around the ship to protect her from the ravages of winter on such an exposed shore. The salvage expert James Bremner (1784–1856), along with his son Alexander, was called in with men, and materials were taken to the site onboard one of the most powerful paddle steamers available, the *Prince of Wales*, built at Neath Abbey Yard in 1842.[9] Brunel could not get there to see the damage for himself until 8 December, and was furious that, despite the work already undertaken, the ship was '…lying unprotected, deserted and abandoned by all those who ought to know her value…'[10] The fact that an iron ship could be refloated after such a disaster focused much attention on the strength of iron over wood; a wooden ship would have broken up in such exposed conditions. However, the tremendous task of refloating and salvaging the *Great Britain* cost the company dear, and the directors were forced to sell the *Great Western* in April 1847 to the West India Royal Mail Co., and the Great Western Steamship Co. itself would have to be wound up, but not before the *Great Britain* was sold. Following an attempt to auction the ship in 1848, the *Great Britain* was finally sold in December 1850 for a mere £18,000 to Gibbs, Bright & Co., who would put her on the Australian passage in 1852.

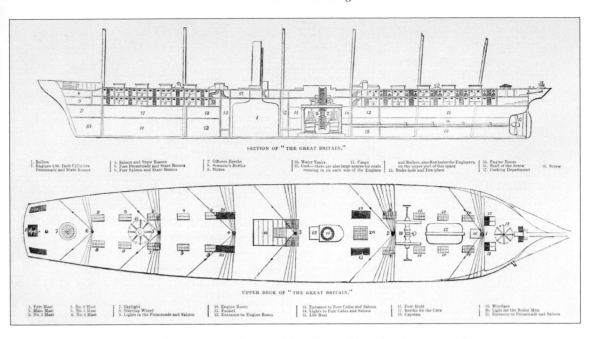

Section and upper-deck of the SS Great Britain. (The Illustrated London News, 15 February 1845)

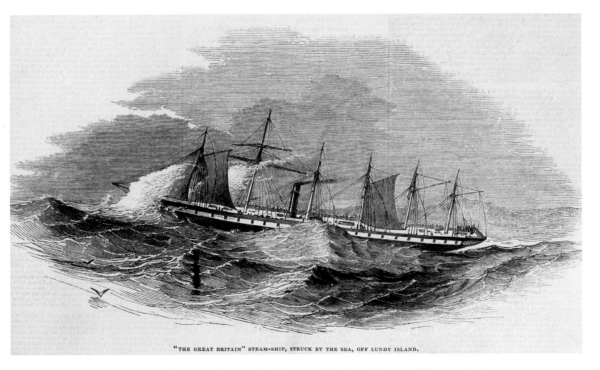

The SS Great Britain off Lundy Island. (The Illustrated London News, 1 February 1845)

William Patterson was called in to convert the ship into a steam-assisted square-rigged sailing ship with four masts, and she was given a new set of engines with twin side-by-side funnels. The Swansea-based *Cambrian* newspaper carried a number of references to a 'Swansea Jack', the first officer onboard the first Australian voyage, one Henry T. Cox.[11] One of the passengers on that voyage referred to him as 'a much superior man', whilst his opinion on the man that had succeeded Hosken, Captain Bernard Matthews, was that he had a 'disagreeable and arbitrary manner'.[12]

The second voyage to Australia of the *Great Britain* began on 11 August 1853, one of the booking agents for this voyage being Henry Bath & Son of Swansea, a company that would figure in the final chapter of the *Great Eastern*.[13] For this voyage she was again rebuilt, this time with three masts and a single funnel. She would make the Australian run her own for twenty-four years until 1876, twice making the passage in sixty-six days, and it is said that one in ten Australians had ancestors who came out in the *Great Britain*.[14] Before taking the story forward to 1886 and the last commercial voyage of the *Great Britain*, there is the story of three further ships that were part of Brunel's steam navigation portfolio. From the time of the completion of the *Great Britain*, Brunel considered the question of employing steam on longer voyages, and was consulted in the early 1850s on the ideal size of a steam ship for the Australia run. The previous decade, from 1840 to 1850, had seen some 520 sailings from Britain to Australia, taking 121 to 130 days on average. In addition to increasing numbers migrating to the new continent, there was also a growing return trade in wool, with Australia supplying 50 per cent of that used in British woollen mills in the 1850s. In the export of wool, specialised companies were taking over from general merchants and using clipper ships to get the wool to Britain, the ships returning with immigrants and general cargo. Seeking to improve the speed and regularity of voyages, the Australian colonies put up a prize of £500 for the fastest steam passage to Australia, and one of the companies interested was the European and Australian Royal Mail Steam Navigation Co. which had been set up to provide a regular mail contract service.[15] William Hawes, the chairman of the company, knew who to consult, being the brother of Benjamin Hawes (1797–1862), who just happened to be Brunel's brother-in-law.

Brunel's recommendations led to the construction of two steamers (fitted with detachable screw propellers) in 1852, the *Victoria* and the *Adelaide*, both 261ft long with a 38ft beam and of 1,350 tons gross. John Scott Russell was the builder and he incorporated his wave-line form into the hull of the ship. The ships achieved a speed under steam of 11.6 knots and a coal consumption of 37 tons a day.[16] The first to be put into service was the *Victoria*, which sailed from Gravesend to Adelaide in sixty-two days, including two days lost coaling at St Vincent, and won the £500 prize. However, even fully booked with passengers and a full cargo both ways, it was reported that she lost £1,000 by her voyage because of the cost of coaling. Being fully rigged sailing ships, the *Victoria* and the *Adelaide* could take full advantage of the trade winds, but both forms of propulsion did not compliment each other as the square rig did not mate with the screw which dragged it back and the masts and rigging hampered steaming. *Victoria* was the most successful of the line and almost halved the average time of the sailing vessels (119 days in 1848), but at the end of her record-breaking voyage the Post Office cancelled the contract.[17] Without the mail subsidy there was no chance of the company being viable but, nevertheless, against the performance of sailing vessels on the same route, they put on an impressive, if short-lived performance. A transitional solution between the age of sail and ships powered solely by steam, these auxiliary steamers were able to overcome the problem of ships held back by unfavourable winds. When wind power was insufficient the screw would be lowered and steam got up, the engines being shut down and the screw being raised when the wind was again capable of filling her sails. Stopping the engines at the right time before speed

"THE GREAT BRITAIN" STEAM-SHIP.—SKETCHED ON THE MORNING AFTER SHE WENT ASHORE AT RATHMULLAN.

The SS Great Britain *as depicted by* The Illustrated London News, *who reported on the 'disaster' when she ran aground in Dundrum Bay off the Irish coast.*

THE "GREAT BRITAIN" STEAM-SHIP, NEWLY RIGGED.

Following the salvage, the Great Britain *underwent refurbishment and was re-rigged for the Australia run. (SKJ collection)*

put an intolerable strain on the screw shaft was an art that had to be learnt by her captain and crew on these new 'auxiliary' steamers.[18]

The need for wind power to supplement steam was based on the fact that an average-size steamship of the day, of 1,800 tons burden (assuming all of this was taken up with the coal required for her journey), would not be able to steam for longer than a journey of thirty-six days burning coal at a rate of 50 tons a day. Therefore, for the Australian run coal depots would ideally be required at St Vincent, the Cape of Good Hope, Mauritius and in Australia for the return journey. There was no source of coal near these depots, so all coal would have to be brought out from Britain by a fleet of colliers.[19] The ships would have to go out of their way to call in to the depots and pay a premium price for the coal, which may be of inconsistent quality, at an average cost of £2 a ton.[20] The example of the loss made by the *Victoria* and the *Adelaide* has already been given, and other steam ships were recorded as losing from £10,000 to £12,000 on voyages to Australia. It was no wonder that fast sailing ships, the famous clippers, were to be developed for this trade. In 1869 the *Thermopylae* equalled the time of the *Victoria* to Australia (but sailing from London to Melbourne).[21] The cargo capacity of such ships, however, was limited, and Brunel considered the prospect of building a vessel that could sail 25,000 miles there and back without having to refuel. If such a ship could maintain a speed of 18mph sailing on the most direct routes, she could make the journey out in between thirty to thirty-five days. Consuming an average of 182 tons a day, 12,000 tons of coal would be required to take her to Australia and back. The ship could take all the coal required at Milford Haven at a cost of 12s per ton, giving a total cost of £7,200 for fuel. Brunel therefore decided that the conventional practice of relying on coaling stations for such a voyage could be abandoned and he began designing a ship big enough to take all the consumables required for a voyage – coal, water and food – with her and back.

Brunel now needed someone to back these ideas and a company that had been incorporated by Royal Charter in 1851 to establish a regular trade route to Australia appeared to be the most obvious choice. This was the Eastern Steam Navigation Co. (ESN) who had attempted to secure a Government subsidy and tendered for a Royal Mail contract.[22] Despite it submitting the lowest tender, the contract, indeed two contracts, were awarded by the Government to P&O in March 1852. With no contract they were open to ideas and Brunel was invited to present his to the board but he was unable to attend and sent Scott Russell, who had built the *Victoria* and *Adelaide* to Brunel's specifications, to present the proposal for a steamship that would be five or six times the size of any then in use. At the scale of dimensions proposed, the vessel could maintain a speed of 15 knots, consuming less coal per ton than an ordinary ship would require in achieving 10 knots. The size would not only allow for a superior level of accommodation and cargo space but enable it to carry enough coal to forgo coaling on the route there and back. A committee reported in favour of the proposition and it was adopted at an ESN board meeting held in July 1852. Brunel was to be the engineer to the company and three tenders would be prepared for the construction of the ship, consisting of the hull, paddle engines and screw engines. However, although the ESN board had accepted the scheme in principle, a number of the directors and the chairman resigned. Henry Thomas Hope was approached to become the new chairman and he and John Yates, the company secretary, managed to fill the vacancies on the board. In August 1853 it was confirmed in the report of the directors that the proposed capital of £800,000, fixed as the amount to be subscribed before contracts would be placed, had been taken up as shares. They were confident that the undertaking was supported by 'men of capital' which would enable the contracts for the size of ships that could be most usefully and profitably employed in the long voyages contemplated, '…whether to India and back or to Australia and back, amount to a voyage round a great circle of the globe…', to be placed.[23]

Tenders had been invited from several parties for the construction of the ship according to the dimensions and power of the vessels determined by Brunel. At that time they had concluded arrangements for the construction of the engines and of the hull, which was referred to as being '…of the first ship' with James Watt & Co., and John Scott Russell and Co., respectively. The ship was to be built on the Thames and was to be completed in eighteen months. The report refers to the large size of the ship as being '…the smallest that will secure the indispensable requisites of comfort, speed and economy.' The report elaborated on the conditions that would be indispensable to regular steam voyages to Australia or India, the first being that they would not be obliged to stop at any place to take on coal. By avoiding this need they were avoiding the delay caused by the time required for coaling and the time lost in deviating from the best route for the voyage to call in the coaling depots. It was stated that existing steamships had lost anything from twelve to twenty days by this practice and were '…incurring the cost of steam without securing its advantages.' The other great saving was that of cost, in that coals on the Indian and Australian route cost on average four or five times as much per ton as in this country. By taking the whole amount of coal for the voyage at a home port it could be obtained for 12s to 14s per ton. An existing steam ship would require 4,000 to 6,000 tons of coal to steam to Australia or India and back; '…the cost of which would supply 15 to 20,000 tons if taken on board at some port in immediate communication with the coal field.' Not only would she carry her own coals but also upwards of 5,000 tons of merchandise and 500 cabins for passengers of the highest class, '…with ample space for troops and lower-class passengers.'[24] The increase in size and magnitude proposed to accommodate all coals required together with the cargo and passenger capacity would greatly increase the factor of speed on the voyage, a velocity of 15 knots an hour, with a smaller power in proportion to their tonnage '… than ordinary vessels now require to make 10 knots.'

This great speed together with an absence of stoppages would, it was proposed, reduce the journey to India by the Cape to between thirty to thirty-five days and to Australia by thirty-three to thirty-six days (the *Adelaide* had taken sixty-two days). The suitability of Australian coal for steaming the ship on the return voyage was raised as an opportunity to reduce the tonnage of coal onboard and increase the cargo and passenger capacity even further.[25] The design of the ship, or ships as the report continued to refer to it in the plural, would be of more than usual strength; the hull, from the bottom up to 6ft above the water line, would be of cellular construction with the upper deck on the same principle, '…so that each ship will be a complete beam, similar to the tube of the Britannia Bridge.' The detail of water-tight compartments and the construction of bulkheads demonstrated the great strength that would be inherent in the ship. The decision about whether the principal route offered should be to India or Australia was not entirely settled. India had initially been seen as the most obvious one just a year or so earlier but that the rapid development of the Australian trade now taking place seemed to be the most remunerative. The direct route for this was considered to be the ocean route to Port Phillip on the south-east coast of Australia, this, the inlet of the Bass Strait, giving access to the ports of Melbourne and Geelong, and also had the advantage of having no 'injurious competition'. [26] By dispensing with costly and inconvenient coaling arrangements, the ESN would not be confined to spending time and resources on these and therefore not compelled to maintain their vessels on those routes because of the investment made. The ESN would be a 'free spirit' in this respect, not bound by subsidy restrictions but enjoying a charter enabling the company '… to apply its capital to any lines which may prove most remunerative…'

There was some reluctance by the directors to give estimates of cargo and passenger charges, but in the case of Australia they had calculated outward charges for goods at the 'very low rate of £4 10s. per ton' with first class passenger rates at £65, second class at £35 and third class at £25, 'including provisions'.[27] On these calculations, and allowing for all expenses, a return of 40 per cent per annum was forecast. They were confident that on these assumptions there would be no need for subsidies, all that was necessary was that which they had acquired from the Government, '…viz. an assurance that no rival route or rival company shall be subsidised…' The forecast of returns of 40 per cent obviously helped to attract shareholders. Brunel, Charles Geach (1808–54) and John Yates had been responsible for much of the lobbying of potential investors to take up shares in ESN, and in December 1853 the first call was made at £3, raising a working capital of £120,000. The list of directors in the 1853 Report included those well-known to Brunel, including St George Burke QC, and C.R.M. Talbot, the latter giving his London address of 3 Cavendish Square.[28] As the chairman of the SWR, Talbot had supported Brunel through the time of the construction of the SWR, a period beset by financial problems and setbacks before final completion to its western terminus at Neyland in 1856. Talbot, and all the other directors and shareholders in the ESN, however, could not have imagined the trials to be faced in progressing the construction, launch and completion of the *Great Eastern*, all of which severely tested the powers of the engineer and was destined to cut short his life, giving rise to and further enhancing the legend that was the curse of the *Great Eastern*. The story of the *Great Eastern* has been told many times and it will be impossible to give the subject the treatment it deserves in just a chapter or two here.[29] The following is an attempt to briefly cover the constructional aspects, emphasising the connection with South Wales, and the later fate of the ship, rather than a continuous narrative of her story.

The *Great Britain* had started life with a different name, albeit before her launch, but the *Great Eastern* had a dual identity, that of *Leviathan*, and she would be launched with that name. In comparing the two vessels, those names are appropriate, the scale of the extinct mammoth compared to the leviathan, the ancient mythical beast of the deep. Brunel, however, had a more affectionate name and referred to it as his 'great babe'. Having been built for the Eastern Steam Navigation Co. (ESN) the name *Leviathan* would not stick to the ship and it would revert back to the *Great Eastern*, and, after all, the name *Leviathan* was stigmatised by the fiasco of the launch. Changing a ship's name was considered unlucky and the *Great Eastern* was regarded by many as a ship cursed by bad luck and misfortune, and that her greatest curse was her size. She would remain unsurpassed in scale for the whole of her life. It was not until 1899 that the *Great Eastern* was exceeded in length with the launch of the White Star Line's *Oceanic II*, and even then not in displacement until the Cunard liner *Lusitania* in 1906. A year later in 1907, fifty years after the *Great Eastern* was launched and some seventeen years after the last pieces of her hull were claimed by the breakers, the *Lusitania*'s sister ship *Mauretania* became the first to exceed her in size overall. The *Great Eastern* would take five years to complete; she had a displacement of 22,500 tons, a length of 6,93ft (211m), a width of 120ft (37m) and a depth of hull of 58ft (18m). Built with a wrought iron double-hull and propelled by both screw and paddle wheels and fully rigged with 6,500 square yards (5,435sq.m) of sail on six masts. The reference in the report to the large size of the ship as actually being the smallest to perform the standards of comfort, speed and economy required is in line with that often expressed by Scott Russell. Giving a paper to the British Association in 1857 on the 'Mechanical Structure of the "Great Eastern" Steam Ship', he sought to correct the impression that he was an advocate for big ships.

The *Great Eastern* was the smallest ship possible for her voyage, claimed Scott Russell, who stated that he was in fact an advocate for little ships. If she answered the purpose for which she was built, '…she would continue to be the smallest ship possible for her voyage.'[30] He also claimed that the big ship was a little short of the proper size, on the basis that her voyage to Australia and back was a journey of 25,000 miles and that her tonnage should therefore be

25,000, but it was actually 22,000 tons. In addressing the British Association, Scott Russell recalled that he had first propounded the form of construction that would become known as the 'wave-line' at their meeting in Dublin twenty-two years ago.[31] From that time to the present he was building vessels, big and small, on the wave-line principle. Around the time that the ESN were considering the construction of a 'monster vessel' (as Scott Russell referred to it), he was building a number of notable wave-line ships. In 1852 Scott Russell constructed the PS *Wave Queen*, an iron ship with a remarkable ratio of length to breadth in the order of more than thirteen times, being 205.7ft long and having a 15.5ft breadth of hull. Her gross register was 196 tons and she was designed for the coastal trade of 100 to 150 miles with a speed of 15.5 knots. 'She was believed to be the smallest vessel capable of the required speed.'[32] A year later he built the PS *Pacific* which also embodied the wave-line principle and was intended to be the fastest ship in the world with a maximum speed of 16 knots.[33] Scott Russell claimed that wherever you find a '…steam vessel with a high reputation for speed, economy of fuel, and good qualities at sea…' you will find a ship constructed on the wave-line principle. With regard to the most obvious quality of the *Great Eastern* – her size – he confirmed that; 'The idea of making a ship large enough to carry her own coals for a voyage to Australia and back again was the idea of a man famous for large ideas – Mr. Brunel.'[34]

Where the ship was to be built was one of the first questions to be addressed. Scott Russell's contract stipulated that construction was to be in a dock for which a price of £8-10,000 was quoted, but this part of the scheme was abandoned due to cost and partly on the choice of a suitable site on which to build the dock. Scott Russell was not happy about the proposal to launch the ship sideways instead of the normal practice of a lengthways and stern first launch.[35] There was little choice, however, because of the length of the vessel entering the River Thames, so that providing the right launch angle would elevate the bow of the ship 40ft in the air. A launch parallel to the river was therefore adopted, to be carried out on a mechanical slip designed by Brunel. This would also be dropped on the grounds of cost, a decision which, with the advantage of hindsight, could be seen as an omen for disaster. A sideways launch required a suitable site and Scott Russell's own yard at Millwall was too small. Adjacent to Scott Russell's yard was the yard that had been established in 1837 by David Napier (1790–1869), another pioneer of steam navigation.[36] He moved to London to be close to the Fairbairn Shipyard and make experiments with steamships. He gave up building steamships there in 1852 as it was, he believed, not the best place to build steamers '…on account of the great expense of production when compared with the North.'[37] With Napier's yard empty, it was possible to lease the yard with materials being moved from the punching machines and plate rollers in Scott Russell's yard by railway. In May 1854 the keel plate was laid, and it was expected that construction of the hull and engines would be completed by October 1855. There was a setback in the supply of wrought-iron plates with the death of Charles Geach on 1 November 1854.[38] On New Year's Day 1855, Scott Russell informed Brunel that he was in financial difficulty and his bankers had refused him further credit. Brunel had to come to a new arrangement with him, but further events over the following year and into the next led to the bank foreclosing on Scott Russell's assets. Scott Russell had failed in terms of carrying out his contract, but what must be put into perspective is the sheer scale of the mammoth undertaking that was the building of the *Great Eastern*.[39]

Scott Russell issued a statement to the ESN repudiating his contract and effectively handing the uncompleted ship back to the company. He had received a total of £292,295 which included extra payments for additional work from ESN, but when progress was reviewed it was found that three quarters of the work on the hull had not been completed. All but £40,000 of the original estimate for the hull had been received by Scott Russell and it was found that some 1,200 tons of iron was supplied but not used on the ship, and could not be accounted for. With work on the *Great Eastern* at a standstill, Yates was prompted by Brunel to negotiate with Scott Russell for the lease of his yard and equipment.[40] The ESN became the shipbuilder and were forced to agree to inflated wages from those previously employed on the ship, as Scott Russell's assistants still held the plans and drawings of the ship, in order to get the work completed. Work restarted in May but progress was painfully slow; at the end of June 1857 Brunel reported that once the screw, screw shaft and sternpost had been installed the ship would be ready for launching by the end of July. However, this would not be on the mechanical slip which Brunel had been forced to abandon and, due to the problems of access to the site, it had not been possible to make the launchways and cradles ready in time. Brunel called in favours, particularly from sympathetic railway contractors like Solomon Tredwell, who agreed to take the contract to build the slipways required for the launch.[41] The lease on Scott Russell's yard had to be extended from 12 August to 10 October but this was still not enough time and on the due date the mortgagees took possession of the yard and blocked entry to all the workmen. Brunel was under great pressure from all sides, and against his better judgement agreed to launch the ship on Tuesday 3 November 1857.

Despite the trials of her launch, at this point in time the *Great Eastern* represented a leap forward in steam navigation technology, but twenty-five years on in 1882 we find the *Great Britain* undergoing a similar leap – backwards. Following the last of her Australian crossings in 1876, and the end of her days as a passenger liner, the *Great Britain* was laid up in the West Float at Birkenhead for several years. She failed to make her reserve price at auction in 1881 but the following year came into the ownership of Antony Gibbs, Son & Co. of London and underwent major and radical changes.[42] Her new owners took up the auctioneer's suggestions that she would make an ideal sailing cargo vessel, carrying coal from Cardiff to San Francisco and returning with wheat. The result was that she was no longer the SS *Great Britain*; her propeller, engine and boilers were taken out and she became a three-masted fully-rigged sailing vessel. Even her iron hull was sheathed with pitched pine cladding.[43] All this took place in 1882, her engines and accommodation being removed at Messrs H. & C. Grayson of Liverpool, a company that would later be involved in the destruction of the *Great Eastern*.[44] The passenger cabins of her liner days were replaced by cargo holds to which access was by three large hatches. To regain balance after her funnel was removed, the mainmast was moved forward. Captain Henry Stap was in command of a cargo ship employed on the run to San Francisco around the Horn, for which she would undertake two round trips, taking her from Liverpool to San Francisco.[45] Stap had taken the commission at very short notice, but he was to regret not joining the ship earlier when he could have been involved in the hiring of the crew, as his opinion of them was low: 'A more useless lot I was never with. Half a dozen of them no sailors at all, substitutes shipped on board at the last moment with only what they stood upright in and are no earthly use.'[46]

He finished this letter to his brother after complaining about the crew, and the anxious time he had, by stating he wished he '…had never seen the GREAT BRITAIN.' A long wait faced the ship on arrival from Liverpool for a cargo and it was six months before she finally secured one. This voyage, the third as a sailing ship, would be the first in which the ship would be loaded with coal in South Wales. The *Great Britain* was now just another one of the many hundreds of ships involved in the South Wales coal trade when she arrived at Penarth to take a shipment destined for Panama on 20 January 1886. There is some discrepancy over the intended destination of what would be the last voyage of the *Great Britain*. Some list it as from Cardiff to San Francisco but she was actually loaded with coal for Panama. Describing her as leaving Cardiff is understandable as Penarth, the actual dock she was loaded from, came under the Port of Cardiff, and the actual destination is academic as she would not reach her scheduled destination. However, a telegram sent from the Falkland Islands confirms, 'The *Great Britain*,

Stap, from Cardiff for Panama…'[47] The cargo that awaited the ship was a consignment of South Wales coal, supplied by Worms, Josse & Co., a French company originally established in Le Havre with a coal exporting office at Cardiff.[48] As a sailing vessel the tonnage of the *Great Britain* was 2,735 gross and 2,640 net, giving her a cargo capacity of nearly 3,000 tons, but her cargo was in excess of this when she was cleared on 5 February 1886 with 3,350 tons of coal. She was loaded from tip No.9 on the Penarth dockside, and one of the tippers who worked on her was Taunton-born John Tidball (1850–1935). Coming to South Wales as a young man, he would work for the rest of his life for the TVR as a tipper at Penarth Dock, recalling the ship's visit many years later to his granddaughter, Elsie Parker-Jones.[49] The voyage was logged as No.47, destination Panama via Cape Horn (the Panama Canal was not opened until 1914), and Captain Stap was the master on what was to be the last commercial voyage of the *Great Britain*.

Like her earlier voyages this was a trip out around Cape Horn, a hazardous voyage against the strength of the prevailing winds. Swansea saw more than its share of 'Cape Horners', men like the twenty-one-year-old Samuel Rees, the mate on the 268-ton sailing ship *Grenfells*, on a voyage out around the Horn to Valparaiso in Chile, taking coal out and returning to Swansea with copper ore in October 1848. He had served his time on his father's schooner *Wave* which early in his apprenticeship onboard had, along with the *Pencalenick* and *Auspicious*, transported the finished chain links for Brunel's Clifton Suspension Bridge from the Copperhouse foundry in Hayle to Bristol.[50] In 1848 a passenger on a similar voyage from Swansea to Valparaiso in Chile recorded his feelings of rounding the horn, with accounts of a gale blowing from the north-west which continued to blow for four days and five nights: 'The ship hove to all this time. The sea was to an alarming height, carried away our other swinging boom.'[51] He believed that he had met his fate more than once; '…but there was but little danger, as we had a fine ship and she has behaved nobly.'

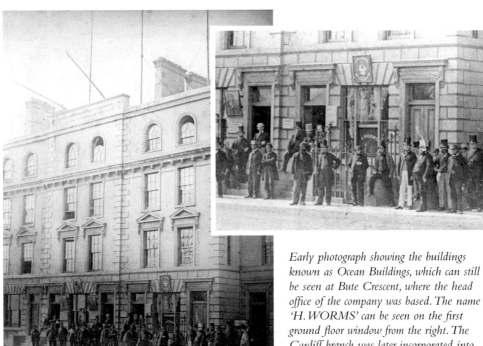

Early photograph showing the buildings known as Ocean Buildings, which can still be seen at Bute Crescent, where the head office of the company was based. The name 'H. WORMS' can be seen on the first ground floor window from the right. The Cardiff branch was later incorporated into the activities of John Cory & Sons Ltd of Cardiff. (Courtesy of Rodney Cadenne)

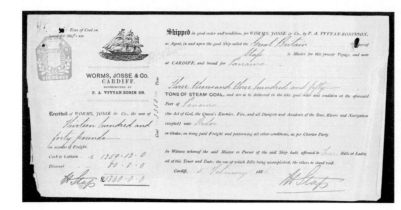

Shipping note for the 3,350 tons of coal supplied by Worms, Josse & Co. to the Great Britain at Penarth. (Courtesy of Falkland Islands Co.)

Telegram of 13 November 1886 confirming the end of the Great Britain as a sea-going vessel. (Courtesy of Falkland Islands Co.)

Family photograph showing John Tidball (1850–1935). A native of Taunton, he came to south Wales and worked all his life for the Taff Vale Railway as a tipper at Penarth Dock. He was to recall the ship's visit many years later to Elsie, his granddaughter. (Courtesy of the late Elsie Parker-Jones)

With no war work to sustain it, Penarth Dock is used for mothballed naval vessels and the sailing vessels Pamir *and* Passat *in this 1950 photograph. (Courtesy of Welsh Assembly Government Aerial Photographs Section)*

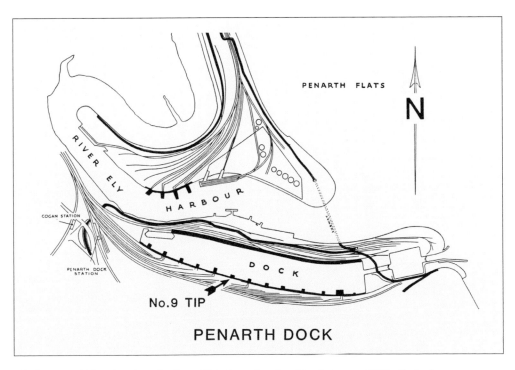

The location of No.9 tip where the Great Britain *took onboard her last cargo. (SKJ collection)*

Great Britain, *seen here on 14 August 1875 at Gravesend, just before her last voyage to Australia. (Courtesy of the National Maritime Museum)*

The Great Britain *in Sparrow Cove. (Courtesy of the Falkland Islands Company)*

Stap, however, did not rate his ship so highly; she was still carrying her original canvas which he felt was now rotten, and much of the rigging was in a similar state. When the ship had to tack against the wind in mountainous seas he expected the mainmast to go at any minute. Voyage No.47 began on the morning tide of 6 February 1886 when she left the safety of Penarth Dock to be pulled by a steam tug past Penarth Head, but she was destined never to arrive at her destination. The voyage was going well until she encountered heavy weather in Latitude 27.25 South, Longitude 43.27 West on 25 March 1886.

It was the start of the difficulties that were to dog the last voyage of the *Great Britain*:

The next day (26 March 1886) a small fire started which was soon got under control but not before some sacks, rope and ship's stores had been destroyed. The following 19 days were plain sailing but on April 16, when in Latitude 54.25 South, Longitude 64.28 West, a sudden south-westerly gale sprang up with the sea breaking right over the vessel and causing the cabin to leak badly. This gale increased in force during the night and by the following night had reached hurricane force. At 8 am on the 18th the crew became nervous and came aft in a body and requested the master to put back to the Falklands, alleging that owing to the ship labouring and straining herself they were afraid to proceed. Apparently the master did not give way, and endeavoured to proceed on the voyage, though on April 21, by which time the gale had moderated, the coal had shifted and gave the ship a list to port; on the 23rd the coal in the 'tween decks was found to be wet. With superhuman efforts these defects were put right, but a few days of moderately good weather gave way to a fresh gale on 3 May which blew continuously for more than a week. The fore and main topgallant masts were both lost on the 10th, and on the 11th another hurricane was blowing. On the morning of May 13, the crew once again came aft and requested the master to put into some port, and this time he decided to do so and chose Port Stanley as he considered it impossible to make Valparaiso. At 3 p.m. on 24 May she arrived off Cape Pembroke, Port William, but her troubles were by no means over, as, while running to the leeward of Williams Islet, she touched ground but came off after about 30 minutes. She then anchored in Port William with 60 fathoms on the port bower, but once again, while swinging to her anchor on May 26 at 4 a.m., she grounded, and this time had to call for the help of the steamer *Rance* to get her off, she was then towed into Port Stanley.[52]

Thus the career of the *Great Britain* was at an end, a ship that began life as the largest steamship of her day and ended as a three-masted barque transporting Welsh coal. Upon being condemned as a constructive total loss, she was purchased by the Falklands Island Co. and used as their main wool store for many years until replaced by the *Fennia* in the 1930s. It had been planned to take her out to sea and sink her, but at the last moment and owing to her historic interest, it was decided instead to take her over to Sparrow Cove where she was beached. Thirty-odd years later she was brought back from the Falklands and can be seen today in the Great Western Steam Ship Co. Dock at Bristol, the original dry dock where she was built and launched in 1843. Sparrow Cove was to be a temporary stay, but at the time it was thought that this was her final resting place: 'Now she rests, a great lady, very old and very frail, but still beautiful, in the land-locked peace of Sparrow Cove.'

Chapter 6 Notes

1 Scott Russell, John (1857), *Mechanical Structure of the 'Great Eastern' Steam Ship*, paper given to the British Association for the Advancement of Science.
2 Rowland, K.T. (1971), p.22, *The Great Britain* (David & Charles: Newton Abbot).

3 *The Examiner*, 4 August 1839. Announcement made by the proprietors of the *Great Western*.

4 *The Bristol Mercury* published a story on 10 June 1843, some weeks after her reported loss, a total loss of the ship and cargo, but only one person, a crew member, is reported to have drowned.

5 *The Newcastle Courant*, 23 September 1842, 'The "Great Britain," or as she is often called, the "Mammoth" steam-ship…'

6 See Chapter 6.

7 Corlett, Ewan (2002), pp.88–90, *The Iron Ship, the story of Brunel's SS Great Britain* (Conway Maritime Press Ltd: London).

8 Corlett, Ewan (2002), p.107.

9 Farr, Grahame (1967), p.95, *West Country Passenger Services* (T. Stephenson & Sons Ltd: Prescot).

10 Corlett, Ewan (2002), p.113.

11 *The Cambrian*, 17 December 1852 (he was the brother of the Governor of Swansea Jail).

12 Fogg, Nicholas (2002), p.44, *The Voyages of the Great Britain; Life at Sea in the World's First Liner* (Chatham Publishing: London). Cox retired following the return voyage and *The Cambrian* was proud to report on his service testimonials in their edition of 25 March 1853.

13 *The Cambrian*, 3 June and 29 July 1853.

14 Smyth, Dacre (1992), p.30, *Immigrant Ships to Australia* (published by Dacre Smyth). The *Great Britain* had a break in 1855 as a Crimean troopship.

15 Kirkaldy, Adam W. (1914 reprinted 1970), pp.78–9, *British Shipping, Its History, Organisation and Importance* (Kegan Paul, Trench, Trubner & Co. Ltd: London, David & Charles Reprints: Newton Abbot). The company is referred to as the Australian Mail Steam Navigation Co. in Ruhen, Olaf (1976), p.70, *Port of Melbourne 1835–1976* (Cassell Australia Ltd: Melbourne).

16 Armstrong, Richard (1974), pp.126–7, *Powered Ships, I: The Beginnings* (Ernest Benn Ltd: London and Tonbridge).

17 Ruhen, Olaf (1976), p.70. The company operated five vessels which ran little more than a year.

18 Another such auxiliary steamer was the *Royal Charter*, which had been purchased for the Eagle Line (Gibbs, Bright & Co.) in order to fill the gap on the Australia run when the *Great Britain* was requisitioned as a Crimean troopship in 1855. Completed by William Patterson, the ship was wrecked in the terrible storm that would become known as the 'Royal Charter storm' off Moelfre on the coast of Anglesey when she returned laden with Australian gold. Charles Dickens visited Moelfre on 30 December 1859 gathering material for his account that appeared in *The Uncommercial Traveller*.

19 The opportunity of Australian coal had yet to be exploited.

20 *A Pictorial History of the Great Eastern Steamship* (no date but 1859), p.1 (W.H. Smith: London). The price of coal at home ports, i.e., Milford Haven, was put at 12s a ton.

21 Kirkaldy, Adam W. (1914 reprinted 1970), p.574, Appendix 1, Record Sailing Ship Voyages. As early as 1854 John Baines made the passage in sixty-three days from Liverpool to Melbourne.

22 Emmerson, George S. (1977), p.19, *The Greatest Iron Ship: S.S. Great Eastern* (David & Charles: Newton Abbot).

23 Eastern Steam Navigation Co. Report of Directors, August 1853 (ICE Library, London). The sum was to be raised in the form of 40,000 shares of £20 each.

24 Eastern Steam Navigation Co. Report of Directors, August 1853 (ICE Library, London).

25 This in fact was further elaborated on; '…should it be hereafter be found that first class steam coal can be put on board in Australia, so as to coal there for the home voyage…'.

26 The state of Victoria in Australia was known as the Port Phillip District of New South Wales before separation and the creation of the new state in 1851.

27 Based on half of the cabin accommodation taken on the outward trip and one-third on the return.

28 St George Burke would later write about his reminiscences of Brunel for Isambard Brunel's biography of his father. Talbot was a shareholder in the SWR, becoming chairman in 1849, and would

become a director of the GWR following the amalgamation of the SWR with the GWR in 1863.

29 Published accounts include; Emmerson, George S. (1977), *The Greatest Iron Ship: SS Great Eastern* (David & Charles: Newton Abbot); Beaver, Patrick (1969), *The Big Ship, Brunel's Great Eastern A Pictorial History* (Hugh Evelyn Ltd: London) and Dugan, James (1953), *The Great Iron Ship* (Hamish Hamilton: London).

30 Scott Russell, John (1857), *Mechanical Structure of the 'Great Eastern' Steam Ship*, a paper given to the British Association for the Advancement of Science Report 1857 (ICE Library, London).

31 Scott Russell believed that the reception of his ideas then and the publication of his idea in the Transactions of the British Association had led to its almost universal adoption.

32 Spratt, H.P. (1968 reprint with additions, to 1949 edition), p.76, *Merchant Steamers and Motor Ships, Part II, Descriptive Catalogue* (Science Museum Reprint Series: London).

33 A ship of 1,469 tons gross register and 245.2ft long with 32ft beam (54ft across paddle boxes), built for the Sydney & Melbourne Steamship Co.

34 Scott Russell, John (1857), p.196.

35 There were problems in the launch of the *Royal Charter*, which also had to be launched sideways.

36 Napier had been responsible for the boiler and castings of Henry Bell's *Comet* in about 1812, and in 1818 Napier had built the first steamboat to establish a service on the Irish mail route between Holyhead and Howth; this was the Glasgow *Rob Roy*, the first sea-trading steamer in the world.

37 Cross-Rudkin, Peter, and Chrimes, Mike, Ed. (2008), p.504, *Biographical Dictionary of Civil Engineers in Great Britain and Ireland, Volume II: 1830–1890* (Thomas Telford Publishing: London).

38 As a director of Samuel Beal & Co. of the Parkgate Ironworks near Rotherham, the company supplying the iron plates for the ship, he had made an arrangement with regard to payment. Geach would take a large part of the payment for these from Scott Russell in the form of ESN shares, Scott Russell himself receiving part payment in ESN shares.

39 Hickman, Keith, 'Brunel's Great Eastern Steamship, The Launch Fiasco – An Investigation', GSIA Journal, 2005.

40 The yard was mortgaged to his banker who now had to be approached. After some weeks of tough negotiating, it was agreed to lease the yard and equipment until 12 August 1857.

41 See Chapter 5 for other Tredwell connections. Rowland Brotherhood would lend Brunel 500 tons of new rails for the launching ways and made much ironwork specially for the launch. Leleux, Sydney A. (1965), p.23, *Brotherhood Engineers* (David & Charles: Newton Abbot)

42 They absorbed Gibbs, Bright & Co. of Liverpool. Corlett, Ewan (2002), p.154.

43 Presumably for protection when lighters came alongside.

44 See Chapter 9.

45 There was an aborted voyage in November 1882 under Captain James Morris when she returned to Liverpool because of leaks. Corlett, Ewan (2002), p.155.

46 Fogg, Nicholas (2002), p.178.

47 Extract of cable sent via Montevideo and transcribed in O'Callaghan, John (1971), p.132, *The Saga of the SS Great Britain* (Rupert Hart-Davis Ltd: London).

48 The Cardiff branch was later incorporated into the activities of John Cory & Sons Ltd of Cardiff, but Worms was still active in France in 1986.

49 Story recounted to the author in the mid-1980s by Elsie Parker Jones of Penarth.

50 This was in March 1843, see Rees, Colin R. (2000), p.47, *Our family of Cape Horners, Volume I: The nineteenth century seafarers and their relations* (Colin R. Rees: Swansea). The links were made up from Dowlais bar iron, see Vol.I, Chapter 5.

51 Rees, Colin R. (2000), p.53, *Our family of Cape Horners, Volume I: The nineteenth century seafarers and their relations* (Colin R. Rees: Swansea). Taken from the transcript of a letter from a passenger, that was addressed to John James of Redruth, in the West Glamorgan Record Office.

52 Garratt, G.L. 'Hulks at Port Stanley', F.I., in *Sea Breezes*, No.30, Vol. V (new series), June 1948.

7

GOD SPEED THE LEVIATHAN
'...CALL HER TOM THUMB IF YOU LIKE'[1]

The photographer Robert Howlett (1830–58) took three photographs framed with the chain-wound checking drum as his background, none of which included George William Lenox and Brunel standing together. Brunel had asked Lenox to 'stand with him' as Brown Lenox had supplied the record size of chain cable for his ship, now wound around the drum. But only Brunel was to be captured or 'hung in chains' by the camera on this occasion, his clothes stained by the mud that was everywhere on site. The strain of the impending launch is evident, but this photograph is now an iconic representation of the engineer, displaying all that was heroic and noble during the Victorian period. With this diversion to his schedule over, Brunel now got back to the herculean task in front of him – launching his 'great babe', his last ship project and the world's largest single piece of engineering. This was the realisation of a journey that had begun with the signing of the final contract for the hull and which was now ready for launching. It had taken almost four years to get to this point, over twice as long as originally estimated, from 22 December 1853 when Scott Russell, a well-established shipbuilder and the originator of the wave-line system, accepted the contract conditions. The standard of the work was to be to the satisfaction of the engineer who would have entire control over the proceedings and the workmanship. Thus the stage was set for what should have been one of the most fruitful industrial collaborations of the Victorian era. Instead it was to result in a scale of conflict only equal to the monster they were creating. By the time the ship was finally launched, Scott Russell's business and reputation was ruined and Brunel's name was being bandied about as a byword for folly and failure.

To launch the ship two launching 'ways' were constructed, part of Tredwell's contract, running from under the fore and aft portions of the ship down to the river at the low water mark

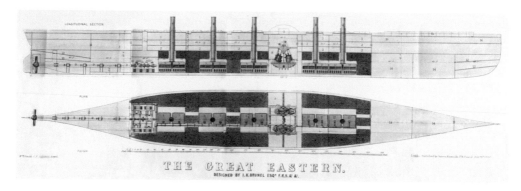

Sections of the PSS Great Eastern *published in 1856 showing the engines, boilers and coaling arrangements. (Courtesy of the Institution of Civil Engineers)*

Right: *Captain Sir Samuel Brown RN (1774–1851) established his first purpose built chainworks at Millwall on the Thames. It was still manufacturing maritime related products such as buoys when this photograph was taken in 1977, but would close in the early 1980s. (SKJ photograph)*

Below: *Detailed map showing the location of the Brown Lenox Wharf and the Napier Yard. Notice also Ferguson's Wharf, the company that supplied the wooden masts. (SKJ collection)*

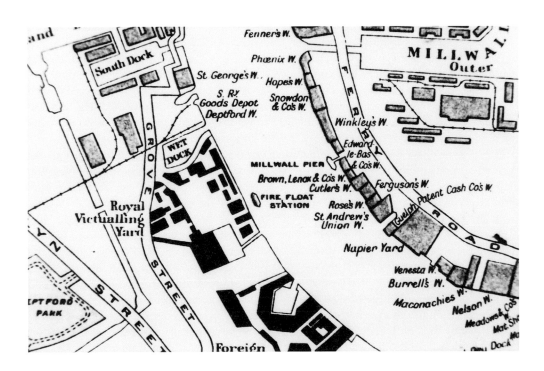

(spring tide).[2] Each of the ways was 300ft long by 120ft wide and the ship was supported by two cradles, each 120ft wide, shod with 1in iron bars. These cradles were 110ft apart and the bow and stern overhung them by 180ft and 150ft respectively. There was no time to test all the various items of equipment before the launch which Brunel had hoped to undertake in an atmosphere of relative peace and quiet, but word had spread and many thousands of spectators manned every vantage point around the yard. In his preparations for the launch he had requested that the men be told to keep quiet during the launch and any verbal orders be given quietly but firmly. In line with his experience of assisting Stephenson at the launch of the *Britannia* bridge tubes and with his own launch of the Chepstow trusses, he would give his orders by means of signal flags from the launching platform. He would be assisted in this by the principal engineer at the works, John Dixon, who had been Scott Russell's engine works manager. To his dismay he found out that 3,000 tickets had been sold by the directors, permitting spectators to enter the yard. As he was getting ready some of the directors joined him on the rostrum with a list of names and asked Brunel which he preferred. His response was 'Call her Tom Thumb if you like', and at 12.30 p.m. Henry Thomas Hope's fourteen-year-old-daughter, Henrietta Adela, came forward and christened the ship *Leviathan*.

The ship may have been christened by Hope but it was not to last long. The men took up their places at 11.00 in the morning, but it was not until after the naming ceremony that the launch attempt began. Brunel ordered that a length of chain cable be eased out from each drum, to which the brakes would be gently applied. A light strain was then put on the two hydraulic presses and a small strain was brought to the tackle at the stem and stern of the ship. He was surprised at how freely the ship then moved when the presses were applied but felt that the pressure on the foremost drum was just sufficient and checked the progress of the ship, but the aftermost drum '…at the heaviest end of the ship the pressure on the breaks [sic] was not sufficient, and the sudden strain upon the chain moved the big drum a little beyond the slack…'[3] This caused the handles of the winches connected to the drum to spin rapidly and throw a number of men into the air, an incident in which five men would be seriously injured with one, John Donovan, dying of his injuries a few days later. Brunel felt that but for this '…accident at the other end of the ship she would have probably gone on slowly.' The

Above: *George William Lenox. The only known likeness is in the form of a bass relief panel. (SKJ photograph courtesy of Brian George)*

Right: *The famous photograph of Brunel against the checking drum, wound with Brown Lenox chain cable. (Courtesy of the Institution of Civil Engineers)*

Opposite: *This illustration has been drawn showing the chain cable without stay-pins or studs (an innovation of Brown's) which maintained the shape of the link and prevented its collapse. It also helped to prevent tangling of the links and was universally adopted for large sizes of chain cable. This, taken from* Pictorial History of the Great Eastern Steam-Ship, *published in 1859, also gives the details of three sizes of chain cable carried by the* Great Eastern, *including a 3in diameter chain cable which is not recorded in the Brown Lenox records (no chain cable manufacturers are mentioned).*

Right: *Directing the first attempted launch on 3 November 1857. From left to right: Scott Russell, Henry Wakefield, Brunel and Solomon Tredwell. (Courtesy of the Institution of Civil Engineers)*

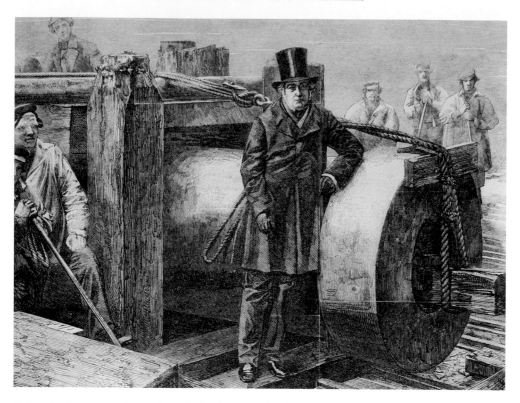

Robert Stephenson standing against a hydraulic ram at the shipyard. (Pictorial History of the Great Eastern Steam-ship, *published in 1859*)

setback caused the launch attempt to be delayed for over two weeks, during which time the inquest into the dead man was held. This was held at the Poplar Hospital, where the workman Donovan died, before the coroner, Mr Baker. The coroner opened the case by stating that the injuries from which the deceased had died were received in an attempt '… to launch the largest vessel ever constructed to swim upon the water – the *Leviathan* – a vessel weighing more than 10,000 tons.'[4] A witness for the deceased stated that he was about seventy-five years of age and that he, George Ross and Donovan, were 'placed' with special instructions for the launch. Ross's instructions were given to him by the foreman, Andrew Ewing, and he would receive the signal to act from a man placed between the drum and the ship, when he would unhook the tackle. As soon as the ship started to move he received the signal and carried it out, he then immediately turned around and saw '…Donovan thrown up by the handle of a winch connected with the drum. The winch went round with great velocity, and Donovan was thrown upwards about 8ft. He fell upon his back, and when I picked him up he was quite insensible.'

   Brunel was asked if he wished to ask the witness any questions, but he declined, saying he was there as a witness and would rather not act as an advocate. When he was examined he identified his role in the proceedings and gave the pertinent facts about the ship, including its dimensions, to which the coroner replied, 'She is the largest ship known in the universe.' No doubt Brunel was used to such comments as he simply responded by stated that she was an iron ship of 10,300 tons. The coroner then asked him to explain why she was being launched sideways instead of stern on. Brunel replied that it was not a new development: 'The American vessels are generally launched broadside on, and at Liverpool the large iron floating vessels are launched in that manner.' He went on to explain the purpose of the machinery being employed in the launch and what had happened during the launch attempt. With regard to the role of the men employed on the winches, Brunel stated that ten men 'properly belonged to these handles', of whom the deceased was not one of them, but that four men were holding the handles at the time, instead of withdrawing as directed. Andrew Ewing confirmed that Donovan was not at his post when the accident occurred and another labourer, Richard Webb, stated that Donovan was partially seated upon the handle at the time of the accident. Other witnesses distinctly heard the engineer in charge of the drum giving the order to 'stand clear'. With the last of the evidence given, the coroner observed that Donovan had been the cause of his own disaster but that it was up to the jury to decide if there had been any want of care or caution on the part of Brunel, or of the persons in charge of the works. The jury returned the verdict of 'Accidental death, caused by the deceased's own imprudence.'

The saga of the launch continued and what followed next was a long war of attrition to reduce the space between the ship and the water. On 19 November another attempt had to be abandoned. Nine days later the ship started moving at a rate of about 1in a minute, but, after a brief stoppage the movement could not be started again, and the rails were compressed into the timbers with the cradles now lying in small hollows. Mooring chains on two of the four midship barges broke and this attempt had to be abandoned, but the ship had moved a further 14ft closer to the Thames. Despite repairing the chains overnight, the same problem occurred on 29 November and, desperate to make real progress, Brunel attempted to borrow as much equipment as he could, and another 8ft was achieved that day. The next day, 30 November, saw movement of about 8½ft before a 10in jack on the forward cradle failed. Brunel applied two extra hydraulic jacks to each cradle and the moorings for the barges were strengthened which succeeded in moving the ship a further 14ft. On 3 December the ship was inched a further 14ft closer to launch and the next day saw another 14ft before two rams, one of 14in and another of 7in, failed and all attempts were abandoned for the day. Throughout this period the press coverage was intense with the letter columns and Brunel's mailbox full of criticism and suggestions for getting the ship launched. In the opinion of *Mechanics Magazine*, '… he made a fatal innovation by introducing iron rails upon his slide-ways, and iron plates upon his bilge ways.' *The Times* reported that the resistance of the iron bars of the cradles on the rails laid on the ways, both surfaces of which were considerably rusted, combined with the great friction caused by running iron on iron, '… offers such a bar to the further progress of the vessel and will require half the hydraulic presses in the kingdom to overcome'.[5] Brunel's decision to use iron as the sliding medium was, in one writer's opinion, 'the greatest technical misjudgement of his career.'[6] *Mechanics Magazine* believed that in spite of all the power so far exerted to get the ship to the water, '… all the King's horses and all the King's men will be required before the task is accomplished.'

It was a task Brunel was resigned to, and he would persevere in his principle of action. Ten years earlier during the struggle to protect the *Great Britain* beached at Dundrum Bay, he had impressed this principle upon Claxton. A principle he had always found to be very successful, sticking obstinately to one plan (until he believed it was wrong); '…and push that to the utmost limits …and if that force first brought up is not sufficient, to bring ten times as much…'.[7] Brunel warned against backing another idea, '…in the hope of finding it easier.' *Mechanics Magazine* makes a reference to Scott Russell, citing him as the originator of the comment circulating that the launch failed '…because the friction of oiled iron on oiled iron is an unknown quantity, the friction of wood on wood with tallow interposed being a known one.' *Mechanics Magazine* referred to the episode of the launch as being '…the greatest and most costly example of professional folly that was ever seen.'[8] The report opened by stating that Brunel was no stranger to failure, but that his failure to launch the *Leviathan* had lowered the reputation of English engineers. He had concentrated the attention of the world upon him with the construction of such an enormous vessel, and what the Thameside spectacle presented to thousands was:

> ... an English engineer at the head of multitudes of mechanics and labourers, breaking ponderous engines, rending enormous cables, crushing solid masses of timber, bursting strong iron vessels, forcing up the soil, tearing up the very bed of the river, expending vast sums of money, impoverishing shareholders, ruining the vessel herself, spreading terror around, imperilling life – keeping this up day after day, week after week, and even month after month, and all in order merely to lower a ship from the shore to the river!

Visitors continued to visit the spectacle that was still, painfully, unfolding. The Princess Royal, the Duchess of Atholl, the Marquis of Stafford and Sir Joseph Paxton were in attendance on 5 December 1857. They witnessed another 14ft of progress but new abutments for the rams had to be built in order to maintain contact with the cradles. Another attempt on 7 December was delayed when the water supply of two of the hydraulic rams failed, but in the afternoon the ship was encouraged to slide a further 8ft. Once again the mooring tackle of the haulage gear could not be relied upon, and on 16 December a further 3ft was gained, but equipment, such as rams and chains, was prone to failure during the course of the day. Brunel was joined at the launch site by his friend Robert Stephenson, when they both agreed that more hydraulic rams or presses were required in order to finish the job. This resulted in the well-known episode with Brunel dispatching one of his assistants to the recently established workshops of Richard Tangye (1833–1906) in Birmingham with an order for hydraulic presses. Tangye would supply the equipment and claimed that, 'We launched the *Great Eastern* and the *Great Eastern* launched us.' Adverse comment of Brunel's technique and the continuing failure of the launch continued to be made. *The Times*, in an article about defective ordnance supplied to the War Department, said: 'It is consolatory, after a series of failures worthy even of Brunel in launching the Leviathan...'[9] The launch attempt was now into the following year, but on 4 January 1858 new abutments were ready for the extra presses, including the 21in one used by Stephenson to raise the *Britannia* bridge. The haulage gear was more securely fixed to the Deptford side of the river, but an accident with a barque running upstream resulted in one of the barges sinking.

Ten feet a day was achieved on 5 and 6 January with the ship being squared up on the ways ready for the next attempt. By 10 January the ship was partially afloat at high tide, and over 11 to 14 January she was gradually pushed down the ways until Brunel had to cease operations to prevent her floating off on the high tide of 19 January. Water had to be pumped into the space between the twin hulls to prevent the ship from launching herself. On 20 January, following high tide, the ship was pushed into position ready for the final launch set for 30 January. However, the weather was too windy on that day and the launch was postponed until the following day. Throughout the night the water ballast was pumped out of the ship, and at 1.30 p.m. on 31 January the ship was finally afloat. Four steam tugs towed the *Leviathan* to the Deptford side of the river where she was to be fitted out.[10] Visitors continued to visit the spectacle that was still painfully unfolding. A visit was made by the Queen and the Prince Consort, accompanied by the King of the Belgians and other royal guests in June 1858. Directors waiting to meet them included Henry Thomas Hope, Sir Joseph Paxton MP, Samuel Baker and St George Burke who, along with John Yates and Captain Harrison, accompanied the party along the full length of the ship, walking on a temporary platform on the iron framing of the deck. The party were on board for almost an hour and glimpsed the engine room through a deck hatchway with an idea of the accommodation given by temporary fittings in the principal saloon. The King of the Belgians was pleased to honour the role of John Yates who he recalled as being one of the first Englishmen he had met when landing at Dover in 1816. It was on this occasion that the directors made it known that they would '…restore to the ship the prestige of its former name, and henceforth the vessel will be known as the *Great Eastern*.' Brunel was not present at this occasion and it was reported that he was seriously indisposed in the south of France.[11]

The launch had cost a staggering £170,000, one third of Brunel's estimate for the whole ship, and it was far from finished as she still had to be fitted out. The Eastern Steam Navigation Co. was close to bankruptcy and to prevent ESN creditors seizing the ship, a new company, the 'Great Ship Company', was formed with a capital of £340,000. The *Great Eastern* was bought for £160,000, leaving sufficient funds for the fitting out and then the ESN went into liquidation.[12] Tenders for the fitting out were invited and two were received, one from Wigram and Lucas for £142,000 and the other from Scott Russell for £125,000. Brunel was absent during the time that the fitting out contract was awarded as he had been told by his doctor to take a long holiday. Almost a year was lost in the completion of the ship, and, despite Scott Russell's history with the ship, his tender was accepted, the terms of the contract being that the work was to be completed in six months to enable the ship's first voyage to America to take place in the summer of 1859. Scott Russell was only responsible for the completion of the paddle and auxiliary engines and supervising the various sub-contractors he had engaged to carry out the variety of tasks involved.[13] To enable the vast quantity of coal to be taken onboard, twenty ports, ten to each side, each measuring 5ft square, were fitted on the lower deck to enable coal wagons to pass inside the hull. In the case of fresh meat, live cows, sheep, chickens, geese, ducks etc., were carried onboard and the necessary pens and cages were provided on deck at the stern of the ship. Captain Harrison, the first captain in charge of the ship, was responsible for determining the rigging of the masts of the ship which, from bow to stern, were named Monday, Tuesday, Wednesday, Thursday, Friday and Saturday. Some accounts state that all the masts were constructed of wrought iron with the exception of the stern mast (Saturday), which was of wood so that it would not influence the compass to be situated on or near it.

However, newspaper accounts refer to there being three iron masts rising 122ft above the upper deck, with a diameter of 3ft 6in for a height of 70ft when they taper to 2ft 6in at the cap.[14] Each mast consisted of two plates, each formed into a half-circle and butt-jointed with internal plates. Internal strengthening was effected by riveting discs of wrought iron reinforced with angle iron inside, and from the keel to the upper deck the masts were encased in a square tube of iron plate. The lower yards of the square-rigged masts were also made of iron and were 126ft long and 2ft 6in diameter at the centre. The Chepstow-based works of Finch & Heath were contracted to supply the wrought-iron masts, made according to their patent. This was the company established by Edward Finch who had come to Chepstow from Liverpool as the successful contractor (then as Finch & Willey) to build Brunel's tubular suspension bridge over the Wye. He was to stay in Chepstow and established a works trading as Finch & Heath. Stays of 7½in wire rope were used on the ship's masts, with the exception of the stern mast which was of hemp. The three wooden masts were supplied by Messrs Ferguson, the site of whose mast house was close to Scott Russell's yard and is now occupied by Ferguson's Wharf.[15] The fitting out was completed by August 1859 with 30 August given as the date of the first voyage, but this was to be put back to 6 September. The destination of the first leg was Weymouth, for which the fares were set at £6 or £10 depending on the choice of cabin. From Weymouth,

Above: *One of the Tangye hydraulic rams with, probably, Richard Tangye (1833–1906) standing in front of it. (Courtesy of the Institution of Civil Engineers)*

Below: *One of the checking drums prior to the launch attempt. (Courtesy of the Institution of Civil Engineers)*

*Start of the launch attempt, the artist's impression probably based on the previous photograph. Note the spectators. (*Pictorial History of the Great Eastern Steam-Ship, *published in 1859 by W.H. Smith & Son: London)*

The Great Eastern *slowly moving towards the river. (Courtesy of the Institution of Civil Engineers)*

Left: *Christopher Rice Mansel Talbot (1803–1890), the Welsh industrialist, SWR chairman, supporter of Brunel's* Great Eastern *and a director of the Eastern Steam Navigation Co. He personally gave Brunel a cheque for £60,000 in order to help with the costs of the launch and no doubt visited the 'sea of mud' that was the launch site of the ship on numerous occasions. He is seen here onboard his steam paddle yacht,* Lynx, *in the late 1880s.* Lynx *was the first ship to enter the South Dock at Swansea (see Vol.II). As a director and later chairman of the SWR he supported the choice of Neyland as home base for the* Great Eastern *and indeed for any ocean-going steamships. (Courtesy of John Vivian Hughes)*

Below: *Almost there, a view of the bows of the* Great Eastern *as it edges slowly towards the river. (Courtesy of the Institution of Civil Engineers)*

a trial trip across the Atlantic would be made with a return to Holyhead which was selected as the departure port for voyages to North America. The choice of home and foreign ports for the *Great Eastern* was to become a moveable feast. Typical of this was the agreement with the Grand Trunk Railway of Canada to use Portland, Maine, as its port of call in the USA. In preparation for this the railway had a special jetty built to accommodate the ship which in the end would never be used. The original schedule had the *Great Eastern* leaving her berth on Tuesday 6 September for the Nore where she would adjust her compasses, then around the Dorset coast to Portland Harbour where she would be open to visitors from 9 to 15 September, to be followed by a trial trip of up to three days. Then the ship would head for Holyhead.

At Holyhead the ship would be open to visitors from 19 to 26 September, before leaving on her maiden voyage across the Atlantic on 30 September, returning from Portland, Maine, on 1 November 1859. All of this would be thrown in the air, just like the tragic event of the forward funnel exploding off Hastings during the first leg of the voyage! Once again, dates would be pushed back as a result, and the departure to the USA was to be on 20 October. Some 150 and 200 passengers boarded the vessel on 5 September for her first voyage but a further delay put back sailing until the following day. At 7.30 a.m. on 7 September 1859, the *Great Eastern* hoisted the Blue Peter and with four steam tugs in attendance, *Victoria* and *Napoleon* at the bow with *Punch* and *Victor* at the stern, moved down the Thames. Extra precautions were taken with the leading tugs having lines attached to provide assistance in an emergency. Attention of the great ship's progress, particularly at the entrance to the West India Docks, created a crowded river which forced the ship to stop for a time. Delays led to the ship being moored at Purfleet for the night and the following day until, accompanied by hundreds of small craft and watched by spectators along the banks, the ship moved out of the Thames and headed for Weymouth. All appeared to be going well until the ship was off Hastings and there was a violent explosion which blew off the forward funnel killing and maiming several stokers and completely wrecking the Grand Saloon. The funnels were fitted with feed water jackets through which water passed on its way to the boilers. This jacket served two purposes; it preheated the boiler water and it helped to reduce the heat in the saloons through which the funnels passed. The two forward funnels had been fitted with stop cocks, and though both were open when the ship sailed both were closed when the accident occurred. One of the engineers on board, realising what had caused the explosion, checked the other stop cock and, finding it closed, opened it and prevented a second explosion.[16]

Fortunately, at the time of the explosion the Grand Saloon was virtually empty, the only person found there by Captain Harrison and the search party was his own daughter who had been shielded from the blast by a bulkhead. A few minutes earlier it had been full of people. However, those in the boiler room were not so fortunate. One stoker, badly scalded, jumped overboard and became entangled in the paddle wheels and was killed, and thirteen men were injured by the boiling water and steam which blew back into the engine room as a result of the explosion. The ship was in no danger of sinking and it was decided to complete the journey to Weymouth. Its arrival with flags at half-mast put paid to the ceremonies that had been planned to greet the ship. Two men, John Boyd and Michael Mahon, died onboard, with three others being taken to Weymouth Hospital where they died the following day.[17] A section of the funnel would be sold to the Weymouth Water Co. and used in the construction of a new reservoir at Sutton Poyntz. Here it was used as a water filter and apart from occasional overhauls when the chamber was drained, it was to spend most of the next 143 years submerged under water.[18] Scott Russell quoted a price of £5,000 and a time of three weeks to carry out the repairs which, in line with his earlier forecasts, was wildly optimistic. To generate revenue it was decided to allow the paying public aboard. Finally, with the repairs complete, the ship set off for its trial run and then on to Holyhead. There were no passengers on board during the trial and those that had paid for the journey had a refund. Holyhead had been previously visited by Captain Harrison for its suitability as a home port and some of the directors in the new company had interests in the London & North Western Railway.[19] The *Great Eastern* weighed anchor at 3.30 p.m. on Saturday 15 October, and she left Portland Harbour at 10.00 a.m. on the Sunday. Captain Harrison ordered the sails to be set and at one point the ship's speed reached 15 knots or 18mph, with an average speed of just under 14½ knots.

Poor visibility saw the *Great Eastern* sail some twenty miles past Holyhead. On the return the captain stopped the screw and used just the paddles, though Scott Russell was against this, and achieved a speed of 7½ knots. Then just the screw was used and a speed of 11 knots was recorded. The trial over, the ship dropped anchor within the breakwater at 3.30 p.m. on the Monday. The ship was opened to visitors while preparations were made for her maiden voyage to America. On the night of 24 October a severe storm developed in the Irish Sea which was to cause considerable damage all around the coastline of western Britain. At Holyhead it caused considerable damage to the breakwater and a number of vessels were sunk around Anglesey including the *Royal Charter*. On the last leg of her voyage from Australia she was within sight of safety but was to be torn to pieces on the rocks with 446 of those onboard losing their lives. The storm would henceforth be known as the 'Royal Charter storm'. Captain Harrison decided to get up steam and, using the paddles and screw, succeeded in keeping the ship out of trouble. After this it was decided to move the ship to Southampton for the winter, and the ship set out from Holyhead on Wednesday 2 November. At Southampton local dignitaries were entertained onboard and subsequently the ship was opened to paying visitors. Cheap excursions to Southampton were run by the South Western Railway for those wishing to visit the ship. Over the next few months the board of the Great Ship Co. went through a bad time with accusations and counter-accusations flying around in all directions. The shareholders were determined to elect a new board to which Daniel, later Sir Daniel, Gooch (1816–89) was elected as a director, and was to act as the engineer to the company. The new board increased the nominal capital of the company by £100,000 in order to meet the requirements of the Board of Trade, and to get other defects put right so that she could sail across the Atlantic at the earliest opportunity. Work costing £20,000 was carried out by Messrs Langley, Penn & Field.

Another blow to the company came on 21 January 1860 when Captain Harrison, Dr Watson, the ship's doctor, Captain Lay, superintendent purser and his nine-year-old son, Ogden, the ship's coxswain, and five of the crew set out from Hythe in one of the ship's boats heading for the docks at Southampton. When the order to drop sail was given it stuck and a sudden gust of wind caught it and overturned the boat resulting in the drowning of Lay's son, the coxswain and Captain Harrison. The curse of the *Great Eastern* had struck again. Harrison was succeeded by Captain John Vine Hall. On Saturday 9 June the *Great Eastern* left Southampton at 2.00 p.m. on a trial trip sailing down Spithead and round the south coast of the Isle of Wight. The ship then headed for the Start Light, which was reached at 1.00 a.m. on Sunday. Due to problems with the paddle engine boilers the ship turned round and headed back along the same route.[20] Amongst those travelling on the maiden voyage of the *Great Eastern* were three directors of the Great Ship Co.; Gooch, accompanied by his wife and son; Henry Daniel, Captain S.T. Carnegie RN and William Barber.[21] Scott Russell stayed at home but his son Norman was onboard. Passengers boarded on 14 June 1860 and preparations were made for the ship two days later, but after visitors had been sent ashore Captain Vine announced he would not be sailing until the following day, 17 June, as the crew were drunk. Gooch saw this as a lack of discipline by the crew and a failure of the captain to maintain it. The route taken by the ship was the more southerly of the regular steamer routes and this also annoyed Gooch as he had wanted the ship to complete the journey in nine days – it actually took ten days and nineteen hours. Although this first transatlantic voyage went without incident it was regarded as a failure in the way of speed. Captain Hall had much experience on the East India route, but had never crossed the Atlantic before.

The ship's arrival at New York created much excitement when its arrival outside the bar was announced via the Sandy Hook Telegraph Station, and at high tide, under the guidance of the pilot, the *Great Eastern* slowly made her way to her berth at Hammond Street. On the way she steamed past the USS *Niagara* dressed overall, once the largest vessel in the world but now, like every other ship in the world, in the shadow of the ship that passed it by. The Cunarder *Asia* fired its guns in salute, every craft available was providing escort and people crowded every vantage point. During the time the *Great Eastern* spent in North America Gooch records that he played the role of a showman in order to 'sell' the ship as a visitor attraction. Despite significant numbers of visitors – on one day the ship received some 18,000 sightseers – it was not enough, even though they raised about £24,000 against overheads of £14,400 for the time she spent there.[22] Being a showman was not a role Gooch relished and he tried to avoid the 'humbug', as he called it, even to the extent of evading New Year photographers who wanted to photograph everyone associated with the running of the ship. The 'humbug' did not last long, particularly after a two-day excursion to Cape May turned into a shambles due to bad management. Subsequent excursions to Hampton Roads and Annapolis received a low take-up due to the bad publicity of the previous excursion. Cyrus W. Field, whose interests would later result in giving the *Great Eastern* her most useful role, was on the Cape May trip and was not impressed with the arrangements. In hindsight it would have been better if the directors had taken up the offer of the world's greatest showman, P.T. Barnum, to run the excursion.

Gooch breathed a sigh of relief when they left New York to sail to Halifax, going there at the express request of that city. However, it was the home of Samuel Cunard and his expanding shipping empire, and on arriving he encountered what he called '…a little sharp practice'. Crowds gathered to see her arrive but the ship was charged lighthouse dues and presented with a bill of £350 as unfortunately for the *Great Eastern* it was based on tonnage! When the Governor of Halifax refused to do anything about this, Gooch was determined not to spend more time there than they needed to. Working all of the night of that day (18 August), they extended the paddle floats out as far as possible as the ship would be very light on the return voyage and therefore ride higher in the water. They were then ready to sail the next morning and there would be no time for sightseers, to which Gooch wrote: '…let the people see the ship in England.'[23] The ship, however, would not return to England but to the place that Brunel himself had inspected as a suitable home port for his 'great babe' just three years earlier; Neyland, in Milford Haven.[24] Milford Haven is an estuary completely different from any other river system in South Wales; technically it is a 'ria', an estuary eroded by a river. The conjoint East and West Cleddau and the estuary has been drowned by the rise in sea level since glacial times.[25] This natural waterway was now the destination for the ship, to which the name *Leviathan* still stuck. Mythical sea creatures or indeed any large sea monster or creature, have been described as 'Leviathans'.[26] Herman Melville used the term in reference to great whales in his novel *Moby-Dick*, and that is yet another reference to Milford Haven.[27] It would play a significant role in the development of whaling and encouraged Nantucket whalers to settle there. Sir Charles Francis Greville (1749–1809) was the moving force behind this, acting on behalf of his uncle, Sir William Hamilton (1731–1803), resulting in at least fifteen Nantucket whalemen and their families coming to Milford Haven in 1792, led by the Starbucks and

The deck of the ship prior to launch. (SKJ collection)

Brunel on board and standing before the forward paddle boiler funnel on 2 September 1859 just before her maiden voyage. It was the last photograph taken of the engineer. (Courtesy of the Institution of Civil Engineers)

CAPTAIN HARRISON, COMMANDER OF THE GREAT SHIP.

Captain William Harrison (1812–60) chosen from about 200 for the position of commander of the Great Eastern. He was appointed in January 1856 and sadly drowned four years later. (Pictorial History of the Great Eastern Steam-Ship, published in 1859 by W.H. Smith & Son: London)

Laying off Deptford after her protracted launch, the Great Eastern *awaits fitting out. (*Pictorial History of the Great Eastern Steam-Ship, *published in 1859 by W.H. Smith & Son: London)*

the Folgers, to establish a whaling port.[28] As early as 1787 Hamilton, when writing about the Milford Haven area, predicted to Greville that the port and the town of Hubbertson would '… one day … be as great as Portsmouth and Plymouth.' Work on the town promised by Greville had not started when the Nantucketers arrived but a speculative new town intended to service the new port and dockyard on the Haven would be laid out to designs by Jean-Louis Barrallier (1751–1834) who, like Marc Brunel, was an exiled Frenchman with royalist sympathies. He escaped the revolutionaries at the siege of Toulon and worked with the British fleet in the Mediterranean where he came into contact with Greville and Hamilton.[29] Admiral Horatio Nelson, visiting the harbour with the Hamiltons, described the harbour of Milford Haven as the next best natural harbour to Trincomalee in the East Indies.[30] He would go further and give what could be seen as the best possible endorsement of Milford Haven, claiming it was '…the finest port in Christendom.'

CHAPTER 7 NOTES

1 Brunel's response to the names for the ship suggested by the directors.
2 *The Times*, 8 September 1857.
3 *The Times*, 10 November 1857.
4 *The Times*, 10 November 1857.
5 *The Times*, Tuesday 17 December 1857, also reported as part of its 17 December 1857 update on the launch, in the *Mechanic's Magazine*, 19 December 1857.
6 Hickman, Keith, 'Great Eastern Steamship, The Launch Fiasco – An Investigation', GSIA Journal, 2005. This paper examines the adoption of iron instead of wooden launchways, which the writer believes could have accomplished a first-time launch.
7 Fogg, Nicholas (2002), p.24, *The Voyages of the Great Britain; Life at Sea in the World's First Liner* (Chatham Publishing: London).
8 *Mechanics Magazine*, 19 December 1857 edition reporting on 'The Launch of the "Leviathan"' on Tuesday, 15 December 1857.

9 *The Times*, 21 December 1857. Ironically the article supports the work of Whitworth in developing polygonal spiral-bored rifle barrels, an area that Brunel had worked on but had withdrawn due to Whitworth's patents.

10 The steam tugs are *Victoria* and *Pride of all Nations* at the bow and *Napoleon* and *Perseverance* at the stern. When a barge fouls the starboard paddle wheel Captain Harrison is forced to order that it is sunk.

11 *Daily News*, 29 June 1858 and *Birmingham Daily Post*, 29 June 1858, re: South of France reference.

12 Existing ESN shareholders were given the market value of their £20 shares, £2 10s (£2.50), towards payment for shares in the new company.

13 See one of the suggested books for a fuller account of cabin accommodation and furnishings.

14 *Glasgow Herald*, 12 July 1859.

15 One of these wooden masts survives at Liverpool FC's Anfield Stadium as a flag pole that sits at the Kop end. It was acquired when the ship was being broken up at the nearby Rock Ferry.

16 The build up to the explosion began when the officer of the watch, Mcfarlane, was on duty in the paddle engine room and was having problems with the donkey engines which pumped water to the boilers. To overcome this he decided to bypass the feed water heaters. These heaters, full of water, were now sealed at both ends and, heated by the funnels, were an accident waiting to happen.

17 The inquest, held at Weymouth, was opened on Monday 13 September and eventually a verdict of 'Accidental Death' was recorded. The five members of the crew who died in the explosion are buried in Weymouth churchyard.

18 Following a major refurbishment by the successors to the water company, Wessex Water, the 7ft-diameter by 5ft-high wrought-iron section was moved to the SS *Great Britain* Museum at Bristol on 14 December 2004.

19 It was planned to offer discounts and reduced fares for *Great Eastern* passengers. Cheap excursions would also be run from the Midlands to see the great ship.

20 The boilers kept priming due to a defect in the wooden casings fitted round the forward funnels which should have dispersed the hot air. They were later replaced with iron lattice work.

21 Henry, Daniel (Harry) Gooch (1841–97) would become the managing director of Bettws-y-Coed slate quarry in 1865. Passengers included: Major Balfour, Mr Beresford, G.D. Brooks, H. Cantan, Mr Cave, Zerah Colburn, Captain Drummond, Mr Field, T. Harnley, Lt-Col Harrison, G. Hawkins, Miss Mary Ann Herbert, A.L. Holley (*New York Times*), Mr Hubbard, D. Kennedy, R. Marson, H. Marin, Captain Morris RN, Captain McKennan RN, Mr McKenzie, Mr Merrifield, Mr Murphy (New York pilot), J.S. Oakford (London Agent, Vanderbilt Line), G.S. Roebuck, Norman S. Russell, Mr Skinner, the Revd Mr Southey, Mr and Mrs Stainthorp, W.T. Stimpson, G.E.M. Taylor, Mr Taylor, General Watkins, H.M. Wells, George Wilkes A. Woods (*The Times*), A. & M. Zuravellov.

22 Dugan, James (1953), p.78, *The Great Iron Ship* (Hamish Hamilton Ltd: London). The exhibitions were, however, a financial failure as the directors were hoping to net £140,000 to offset some of the debt of the company which was incurring a level of interest of £1,000 a day.

23 Wilson, Roger Burdett, Ed. (1972), p.83, *Sir Daniel Gooch: Memoirs and Diary* (David & Charles: Newton Abbot).

24 In July 1857 Brunel visited Neyland and Milford Haven; the *Haverfordwest & Milford Haven Telegraph* speculated that his visit was in connection with settling a home port location for the *Great Eastern*. See Vol.II, p.169. In July 1857 Brunel visited Neyland and Milford Haven; the Haverfordwest & Milford Haven Telegraph speculated that his visit was in connection with settling a home port for the *Great Eastern*. See Vol.II, p.169. It was also selected by Shakespeare, who wrote *Cymbeline* in 1611. In Act III, Scene 2 Milford Haven is referenced in the lines by Cymbeline's daughter Imogen, who asks:

'To the smothering of the sense—how far it is
To this same blessed Milford: and by the way
Tell me how Wales was made so happy as
To inherit such a haven: but first of all'

25 Bird, James (1963), pp.211–212, *The Major Seaports of the United Kingdom* (Hutchinson: London).

26 Leviathan was used by Thomas Hobbes as the title of his celebrated work describing the state as a vast living organism.

27 When *Moby Dick* was made into a film starring Gregory Peck, the director John Huston chose Milford Haven for much of the location filming because of the rich whaling heritage there.

28 Historic Nantucket, pp.4–7, Vol. 56, No. 1 (winter 2007).

29 Barrallier laid out the town in a square pattern which can still be seen today, including St Katherine's church, built between 1802 and 1808 to a design by Barrallier or his son Charles.

30 Constantine, David (2002), pp.274–275, *Fields of Fire: A Life of Sir William Hamilton* (Pheonix Press: London).

8

BLESSED HAVEN

'…THE HIGHWAY TO AMERICA. (HEAR, HEAR)'[1]

Helegovians never got the opportunity for close viewing of the spectacle that was the *Great Eastern* for at 9.00 a.m. on Sunday 19 August 1860 the ship weighed anchor at Halifax and returned across the Atlantic. Just seventy-two passengers witnessed a new eastbound speed record of nine days and four hours, however the earlier supposition that there was a build-up of marine growth on the ship's hull from her time at Southampton was felt to be the reason that even greater speeds were not achieved, the ship's average speed on the return journey being between 13 and 14 knots. This, together with a suspected problem with the propeller shaft bearings, made it necessary to visit the western terminus and port of the South Wales Railway of Neyland at Milford Haven.[2] This was the embryonic port that Brunel had originally selected, and which C.R.M. Talbot was keen to support, as the home base for the *Great Eastern* and indeed for any ocean-going steamships. At 4.00 p.m.on 26 August the *Great Eastern* dropped anchor at this safe haven, Daniel Gooch recalling; 'It was a very grand sight as we steamed up Milford Haven.'[3] The Channel Fleet, some eleven or twelve ships strong, was lying in line along the Haven, and as the great ship passed these wooden walls they manned their yards and gave hearty cheers. A special train was waiting to take the passengers on to London but, unusually for the completion of any transatlantic voyage, no one wanted to leave the ship just yet. The passengers elected to spend another night onboard and the departure was postponed until the following morning when, along with the other passengers, Gooch left on the special train to make his way back home to Clewer Park, near Windsor, a journey that could be made in about eight-and-a-half hours, assuming the train stopped for him at Windsor. He noted; '…such is the comfort of the *Gt Eastern* that all felt regret that it was time to leave her.'[4]

Gooch admitted in his diary that the voyage was not a profitable one and operating expenses had taken up all the money they had made by exhibiting her. However, there was some consolation in that no defects had shown themselves in the operation of the machinery. In the light of what should have been obvious to many and certainly to Gooch, that the ship was too big to be profitable on the Atlantic run, why wasn't the ship being used on the run it was designed for – to Australia or to India? The Great Ship Co. had only one ship, so they could not provide the scheduled two-way service necessary for commercial success on the Atlantic run to be profitable. On such a run the ship would have to capture most of the current passenger traffic crossing between Britain and the USA to fill her 4,000-berth capacity – traffic it would have to take away from steamships specifically built for the transatlantic trade – and her average speed of 13-14 knots was matched by her fastest competitors on that run such as the Cunard ships *Persia* (1856) and *Scotia* (1862). The *Persia* was Cunard's first iron ship and the largest steamship in the world until the launch of the *Leviathan*.[5] Was there a conspiracy by vested interests to keep the *Great Eastern* away from the routes she was built for? William Hawes thought so. Hawes had been the chairman of the European & Australian Royal Mail

In June 1857 a specially constructed pontoon, 150ft long, 42ft wide and 16ft deep, was launched at Neyland. It was to form part of the landing stage designed by Brunel for the Great Eastern *and constructed by 'Mr. Okeden, the resident engineer at Neyland.' Additional pontoons from the floating out of the Saltash tubes were used to extend this pontoon to 500ft. They were protected against the ebb and flow of the Haven by substantial wooden dolphins. See colour section. (*Pictorial History of the* Great Eastern Steam-Ship, *published in 1859 by W.H. Smith & Son: London)*

Steam Navigation Co. when Brunel's advice had been sought on the design of steam ships, leading to the launching of the *Victoria* and the *Adelaide* in 1852.[6] As a shareholder in the *Great Eastern*, he questioned the logic of not using the ship on routes to Australia or India, and would later claim at a shareholders meeting in 1863 that placing her on routes she was ill-equipped for was the reason the ship had financial trouble. However, several more Atlantic trips were destined to be run by the *Great Eastern*, captained by an ever increasing turnover of masters.

No doubt C.R.M. Talbot was hopeful that when the *Great Eastern* was built it would use Neyland as its home port. In February 1853 he reported on the possibility of a dock investment being made on the south and north side of Neyland. This he believed would not only justify the selection of the western terminus at Neyland (as the terminus for the SWR) but also lead to the extension of '…the railway to those docks, and thus making it the highway to America. (Hear, hear.)'[7] *The Cambrian* had reported on the Channel Fleet and the *Great Eastern* and it was proud to mention, in its edition of 14 September 1860, that the chief officer of the ship while in Milford Haven was Mr W.H. Davies, a native of Swansea.[8] At least one man onboard would have been familiar with Milford Haven from service on another Brunel steamship: this was Alexander 'Mack' Maclennan, the archetype Scottish chief engineer who had sailed on the *Great Western* in 1838 and later spent seven years on the *Great Britain* as chief engineer. The Great Ship Co. could not afford to maintain a full-time shore organisation or a full crew complement when the ship was not in service. This resulted in many of the crew, including the Captain, being laid off at Neyland once the difficult manoeuvre of putting the ship on the gridiron was accomplished.[9] Beaching the *Great Eastern* on to the gridiron at Neyland was something that required great skill and seamanship. To quote *The Times*, it was carried out '… with masterly skill, under circumstances too, of the most adverse nature.'[10]

Just before noon on Sunday 16 September 1860 the *Great Eastern* was observed getting up steam for such a manoeuvre. On the bridge could be seen Captain John Vine Hall, Robert

Left: *RAF reconnaissance photograph of Neyland taken on 22 April 1945. The station complex can be clearly seen alongside the River Cleddau as it flows into Milford Haven. Also visible are those 'flying leviathans', the Sunderland Flying Boats based at Milford Haven during the Second World War. (Courtesy of Welsh Assembly Government Aerial Photographs Section)*

Below: *Neyland in the 1970s, seen from the Cleddau Bridge. (SKJ photograph)*

Designed by William Owen of Haverfordwest and completed in April 1858, the impressive South Wales Hotel was the result of a determination on the part of the SWR directors to provide first class facilities for passengers using Neyland. Sadly it was demolished in recent years. (Simon Hancock collection)

Services to Waterford and Cork were run from Neyland by Captain Robert Ford and Captain Thomas Jackson as the Milford Haven and Waterford Steam Ship Co. Malakhoff had been built by Scott Russell in 1851 as the Baron Osy, *with service in the Crimea resulting in a new name. She was employed by Ford & Jackson from 1856 until 1872. Scott Russell regarded her design as the ideal cross-Channel vessel. (A* Visual History of Modern Britain: Transport, *ed. Jack Simmons, Vista Books, 1962)*

Pearson Brereton, Brunel's former assistant who was now acting as engineer to the company, Captain Jackson, the company's agent and Mr Ivemy, the Queen's harbour pilot.[11] The managing director of the company, Thomas Bold, was also onboard along with J.P.G. Appold, who had been present beside Brunel at all of the launching attempts. It has already been mentioned in Chapter 1 that Appold could have been the inspiration for Dickens when describing all the various ingenious contrivances made by Pip in *Great Expectations*, and equipment designed by Appold of an ingenious nature, such as his cable brake, would be employed on the *Great Eastern*.[12] Meanwhile steam was being raised for the ship's paddle wheels and from time to time these made the great vessel turn ahead slowly as the wind rose in order to ease the strain on the single cable she rode at. Around 4.00 p.m. a specially chartered tug took hold of a hawser from the ship, which then began to heave in her own anchor, the tug keeping her in position until it was time to head for the gridiron. When it was judged that the tide was sufficiently high for the manoeuvre, the *Great Eastern* was allowed to drop slowly down, stern foremost with the wind and tide, at a speed of 3 knots per hour. This was checked by her paddles being occasionally reversed while the tug held her head in the required position. An hour later at 5 p.m. the ship was approaching the gridiron, having travelled a mile and a half from her moorings in about twenty-five minutes. When secured on the gridiron access to the ship was to be effected by a timber gangway from the shore that connected with a hatchway near the stern. This delicate structure was seen as being at risk during the beaching manoeuvre, and it was ordered to be cleared of everyone who was not engaged in the operation at hand. While this was being done the ship turned ahead until she was within twelve or fifteen fathoms of the intended position. After some manoeuvring due to the great force of the current, it was decided to wait for slack water and her port bower-anchor was let go and her stern allowed gradually to sheer in.

The *Great Eastern* was pushed by the tide against the eastern dolphin of the gridiron, a force it resisted, and indeed the mass of the ship recoiled two or three times from the uprights, or

The later pontoon at Neyland from which a car ferry ran from here to Hobbs Point, Pembroke Dock. In this early 1970s photograph the new Cleddau Bridge can be seen under construction across the water. (SKJ photograph)

Coaling appliance alongside the station yard, one of the facilities designed by Brunel for Neyland, seen here in a GWR photograph taken in the 1930s. (Courtesy of National Museums and Galleries of Wales, WI&MM collection)

General view of the station showing the goods yard in 1910. The coaling appliance can be seen on the right. (Simon Hancock collection)

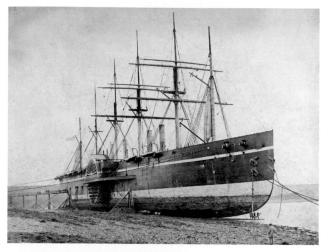

Opposite: *The* Great Eastern *on the gridiron at Neyland with the tide under the hull. This is one of a superb series of photographs taken shortly after the ship was beached on th gridiron in September 1860. (Courtesy of the Institution of Civil Engineers)*

Left: *Tide going out, men at work under the hull. The timber gangway from the shore that connected with a hatchway near the stern can be seen, a delicate structure that had been ordered to be cleared manoeuvring to place the ship. (Courtesy of the Institution of Civil Engineers)*

Close up showing some of the men around the ships Trotman anchor. In the background one of the distinctive Pembroke Dock buildings can be seen across the Haven. (Courtesy of the Institution of Civil Engineers)

'dolphin', to a distance of 6 to 8in. In the meantime a flotilla of boats including lighters and a ferry steamer passed up hawsers and ropes in order to haul up the chains from the four anchors that would secure her on the gridiron. By 6.30 sufficient slack water permitted the final manoeuvre: '…orders were, therefore, given to "turn about," and in two minutes the *Great Eastern* was in her place to an inch!' The gangway was now immediately in front of the entrance in the hull which gave access to the lower deck. Slack was hauled in from the mooring chains to keep her in an upright position and 'snug', moorings consisting of one anchor right ahead with one astern, another on the port bow and the fourth on the starboard quarter. Brereton then ordered all the boilers to be pumped full – what is termed to 'scuttle' them – so that the weight of an additional 80 tons of water would help to keep the ship steady. The site of the gridiron consisted of two timber grids with 'dolphins' on the landward side. The western dolphin was just ahead of the starboard sponson and the eastern one under the starboard quarter.[13] Some thought the use of ballast water had led to damage when the tide receded as this allowed the bow to remain unsupported while overhanging the gridiron with the extra weight of the ballast water still in her.[14] The *Great Eastern* only needed support at

two points along the length of her hull as Brunel had designed the ship in such a way that the entire vessel acted as one continuous iron beam. The gridiron supported the ship at the same points as the launching cradles at the ill-fated launch:

> In constructing the foundation of the floor on which the ship is being built, provision is made at two points, to insure sufficient strength to bear the whole weight of the ship when completed. At those two points, when the launching has to be effected, two cradles will be introduced, and the entire fabric will be lowered down gradually to low water mark, whence, on the ensuing tide, the vessel will be floated off.[15]

Although the ship now appeared secure, the state of the weather still threatened her safety, and steam was kept up until the following morning. The task of scraping and painting the bottom of the ship's hull was planned to be completed by the Tuesday. *The Times* thought that the time allowed '…is too short and the undertaking too vast.' The reporter praised the way the ship preformed in such bad weather, and the capabilities of Milford Haven as a fitting port for the great ship and, showing her magnificence 'As she now lies, with her whole vast size displayed from keel to truck, she forms a noble monument…' – a monument which the reporter reminded his readers with the last words of his story, was '…to the memory of I.K. Brunel, as he may be said to have given his own life to render his darling project a success.' No problems were detected in terms of fouling of the ship's bottom, and it was given a coat of paint, this being reported in *The Cambrian* for 21 September 1860, and some of this work can be seen

Above: *A 'Leviathan'-like image of* the Great Eastern *devouring the boat and throwing the crew into the water. The attempted beaching of the ship on the Neyland gridiron in 1862. (SKJ collection)*

Left: *Sir Daniel Gooch in late middle age. (SKJ collection)*

actually going on in one of the photographs taken at the time. Other painting work included the funnels which were painted bright red and the white line running around the hull and its paddleboxes, much to the annoyance of those who found it gave the ship the optical illusion of a hogged deck.[16] The original timber deck, which was laid on the iron deck, had not stood up well to wear and tear so it was decided to cover it with pitch and put a new 2in layer of pine planking on top. The stern bearing of the propeller shaft was examined and it was found that the special lining of Babbitt metal had been squeezed out of the bottom of the bearing. A new bearing of *lignum vitae* within a brass collar was shrunk on the screw shaft under difficult conditions, but nevertheless proved to be effective.[17] In all everything that needed repairing was done, including damage to her paddle boxes and paddlewheels when she berthed in New York, caused by coming too close to the wharf, damage perhaps not unsurprising as up until then she had always been moored offshore, New York being the first opportunity to berth the great ship.

During the excursion made by the *Great Eastern* from New York to Annapolis, President James Buchanan (1791–1868) visited the ship and brought a large party of his cabinet with him. Buchanan was in office when the second, and successful, Atlantic telegraph cable was completed two years earlier in 1858, and he responded to the congratulations sent by Queen Victoria down the wire. This was the second time that Gooch had met Buchanan on that trip and they discussed American trade, Gooch seeking his opinion on what he thought of the idea of the ship sailing to the southern states and returning to Britain with cotton. Gooch records that 'The President thought well of the scheme.'[18] Like the brief interlude of the 1858 Atlantic cable, which only managed to transmit messages for three weeks, the president's optimism was short lived. Back in Britain the following year it now appeared inevitable that the southern states would secede from the Union, and the Great Ship Co. decided to play safe with a second voyage to New York on 1 May 1861 being advertised in *The Cambrian*.[19] The same newspaper announced in its 22 March 1861 edition that the *Great Eastern* would be taken off the gridiron, which took place on 26 March 1861, and she 'steamed off admirably'.[20]

The voyage was to be under another new man at the helm, but the Hon. S.T. Carnegie, RN (a director of the Great Ship Co.) was destined to have the shortest term of duty of any captain. Prior to sailing, a warrant was instigated against the ship by Scott Russell, seeking to claim the award made to him following arbitration over his repair of the ship and funnel after the feed heater explosion. The warrant was posted by the Sheriff of Pembrokeshire, George Lort Phillips MP, and attached to the ship on 25 April 1861.[21] The company appealed against this but lost, and, with only a few days to go before the planned sailing, reluctantly paid up by taking the money from the cash in hand allocated for the voyage. In order to keep within what remained of the budget, the company were forced to lay off six of the ten senior officers and a third of the crew. The conclusion of this episode marked Scott Russell's last connection with the Great Ship Co. and with the *Leviathan* itself, over a decade after the idea of the ship had first been discussed by Brunel and Scott Russell. It may have been the end of the shipbuilder's links with 'the largest ship known in the universe' but it was not the end of Scott Russell's connections with South Wales. In the meantime Captain Carnegie was refusing to sail under such short-handed conditions and resigned. A last minute replacement was found in the shape of Captain William Thompson along with Mr Robertson as the new chief engineer.

Finally the ship set sail on its second Atlantic trip to the USA carrying 7,000 tons of coal but very few passengers. When about four days out she ran into the worst storm she had so far met but arrived safely at New York. The crowds were no longer in attendance to witness her arrival. After all, there were other things to occupy the attention of New Yorkers; that of the Civil War. Talk of war raised the possibility of the *Great Eastern* being used for military purposes, even that Lincoln's administration might charter the ship for such services.[22] Other rumours being picked up by the British Government included one that the USA might declare war on Great Britain and France in order to bring the southern states back into the Union, and if that happened Canada would be vulnerable. Gooch recorded that this voyage was a good one and more than covered their expenses, bringing back 7,000 tons of cargo, mostly corn, on the return trip to Liverpool. Back home, the War Office decided to charter the ship to carry troops to Quebec to counter any possible threats, including the aspirations of Irish Fenians in the USA, made to Canada. She was to be a war transport, a role that the *Great Western* was engaged in at the end of her career when she was sent to the Crimea, a theatre of operation that also involved the *Great Britain* which would also serve as a troop carrier during the Indian Mutiny. The financial rewards of such a contract brought great rejoicing from the shareholders and she would finally be carrying numbers up to her design capacity. There were 2,144 officers and men along with 473 women and children, together with their equipment and 122 horses occupying a ship hastily adapted by contractors from the Birkenhead Iron Works. In addition, forty independent passengers were also going along, including Henry Marc Brunel and William Froude (1810–1879). Captain James Kennedy, on loan from the Inman line, was another new skipper, in line with the *Great Eastern* tradition. The contract, and her spell as a war transport, was terminated with her return to Liverpool.

For her last voyage of the year she left Liverpool for the second time and headed for New York on 10 September 1861 carrying a record 400 paying passengers. At the helm was another new captain, a former Cunard officer by the name of James Walker, but there would be no honeymoon period and two days out the *Great Eastern* steamed into a heavy gale. The gale rose to hurricane force and she found herself in the trough of mountainous waves that pounded the ship and caused much damage by the uncontrolled movement of cables, lifeboats working loose on their davits and shifting cargo. The destructive force of the elements was compounded by incorrect handling of the ship during attempts by Walker to bring the ship into the wind to escape the valley and meet the waves head on. At the time Gooch was staying with the chairman of the Great Ship Co., Samuel Baker, at his house at Thorngrove, near Worcester, to discuss the future of the ship.[23] Until the first telegram arrived informing them that the ship was now off the Irish coast they thought fortune had at last turned and that the future of the ship was now a bright one. Sadly this was not the case and, as the ship at this time should have been three days away from New York, they knew something had gone disastrously wrong and she had turned off course. The ship was towed into Cork Harbour on 17 September 1861 with most of the passengers getting off, believing themselves lucky to be alive, and all were offered refunds. Gooch simplified the catalogue of misfortune and misery encountered on this voyage in his diary, writing that her rudder shaft had broken and that she had been at the mercy of the elements for several days until temporary arrangements were made to steer her:

> She had a very bad time of it, but got safely into Queenstown. Both her paddles were carried away. We got her over to Milford again and put her on the gridiron for repair; these repairs were very costly and cleared away much more than our profits.[24]

The following month the *Great Eastern* was escorted by tug to Milford Haven and rumours circulated that the ship was being sold to the Emperor of France, and on 25 October *The Cambrian* reported that her coals were being advertised for sale.[25] Over the winter of 1861–82 she underwent repairs at Neyland including the installation of replacement paddlewheels 52ft in diameter which were smaller (6ft less in diameter) than the originals but stronger, and a new steering gear that could be operated in emergency from two lower stations. The paddlewheels were fabricated at the Chippenham Railway Works of Rowland Brotherhood, who sent his son, Peter, down to Neyland with a gang of men to fit them to the ship. Brotherhood also fabricated a rudder post of riveted iron weighing 8 tons.[26] The second grounding of the *Great Eastern* on the gridiron would not be as straightforward as the first, as on the February 1862 occasion a 'most deplorable accident' occurred. A boat from HMS *Blenheim* was assisting and cables were thrown from the *Great Eastern* to the *Blenheim* boat. While the hawser was being paid out, the tide caused the slack of the hawser to drift under the screw; '…and the first revolution drew the *Blenheim* boat right in on the screw.'[27] Of the men onboard, thirteen in number, nine threw themselves into the water but the remaining four failed to escape in time and the boat was rapidly sucked into the maelstrom formed by the screw revolution:

> All hope seemed to have abandoned them, when one of the fans [propeller blades] threw the boat up, and then drew it in between the screw and the vessel. The accident was so unexpected and so sudden, that it was impossible even to cut the rope before the boat was sucked in upon the screw. The four men were speedily rescued from their dangerous position, and the screw was then gently moved, in order, if possible, to let the boat down uninjured. The first rise of the fan, however, smashed her in atoms, and the pieces of her wreck were whirled in all directions by the rapidly incoming tide.

Meanwhile the men who had jumped into the water were in peril due to the rush of the tide and two drowned; it was noted that a ferryman passed the scene only a few yards away but made no effort to save the drowning men: '… the worthless and heartless wretch paid not the slightest heed…' The sight of this must have presented an image of the 'Leviathan' similar to the image feared by sailors as a gigantic whale-like or sea serpent monster that devoured whole ships by swimming around the vessels at such a speed that it would create a whirlpool. After this, but before she was safely moored, she crashed into HMS *Blenheim*, an accident blamed by Gooch on the blundering of a pilot, the company having to pay £350 for the damage to the warship's bowsprit, mainyard and moorings.[28] Another attempt the following day went without incident. By 18 April 1862 it was reported that the repairs had been completed, repairs that cost £60,000 which, as Gooch had remarked, cleared away much more than their profits.

The next voyage would again be to New York from Neyland, and she was delayed from sailing on 6 May due to her hawsers being fouled, but the following day, under yet another new captain, Walter Paton, she steamed out of Milford Haven. Passenger numbers were again disappointing outbound but the return trip was more successful. It was remarked upon in the New York papers that 'Captain Paton's mates and other subordinates are all different from those who first came over on the ship'. They, however, noted that 'The only one of prominence still retained is Chief Engineer McLennan.'[29] Things were looking up as Gooch remarked that the ship made three voyages that year, '…each time increasing the number of her passengers. The last voyage, in August, she carried 1,530 out, but unfortunately…' with the luck of the ship it was almost inevitable that such a run of successful voyages would be interrupted. On arriving at the approaches to New York off Montauk Point soon after midnight on 26 August 1862 she heeled a few degrees to port but her progress was unaffected.

The pilot had just come aboard and he suggested that they had hit an uncharted reef called the 'North-east Ripps', which had sliced through the outer hull near the bottom of the ship. One hole was 80ft long by 6ft wide and there were about eight smaller holes.[30] This reef would henceforth be known as the 'The Great Eastern Rock' on the charts until it was finally removed. Such unchartered rocks were not uncommon; indeed in 1860 Admiral Fitzroy remarked on a rock that had lately been discovered at the entrance of Milford Haven: '…in the fairway channel, upon which it was possible for a large ship to strike.'[31] At New York the inner hull was intact and the ship was in no danger of sinking yet it had survived damage that would have sunk any other ship. Captain Paton, in command of the ship on all three voyages that year, a first for the *Great Eastern*, decided to try and make repairs before returning to Milford Haven. This was carried out by the New York engineer Edward Renwick, and detained the ship until January 1863. The work cost £70,000 and once again the company's profits and more were wiped out. Gooch recorded that it '…pretty well ruined us.'

This was, to say the least, a bad start to the year, and a further voyage was undertaken from Liverpool to New York in May 1863, but another storm, in which the ship was buffeted in a trough of waves (echoing the disastrous voyage two years previously), saw the ship losing a paddle wheel. Screw propulsion enabled her to complete the voyage but repairs delayed her return until September. The company were in the red to the tune of £18,979 and the total debt, including bonds such as mortgages and debentures against the ship, amounted to about £142,350, all of which was too much for the directors who finally conceded defeat.[32] William Hawes was appointed as chairman of the shareholders committee, a man who had long criticised the use of the ship on the transatlantic run, and his committee explored various fundraising schemes including a lottery, to finally put her on the Australia run. These came to nought and the ship was put up for auction on 14 January 1864 but failed to meet the minimum reserve necessary to pay off all judgement claims against the ship, and it was withdrawn. The then chairman of the Great Ship Co., William Barber, along with the contractor Thomas Brassey and Gooch, held between them more than half the outstanding bonds on the ship. Gooch had been considering his position with the Great Western Railway at this time and despite his outside interests and directorships with other companies such as the Great Ship Co., he was still the GWR's locomotive superintendent. In 1861 he had received much criticism from the South Wales Railway shareholders (Talbot leading the criticism of his management in this area) over the rates charged by the GWR to provide their motive power. The amalgamation with the SWR and West Midland Railways proposed in 1863 would result in new directors coming on to the new GWR board, of which he would write; '…those likely to come from the West Midland not standing very high in character.' Two of these directors, Richard Potter and Thomas Brown, had moved to sell the Gyfeillon colliery in the Rhondda, a colliery that Gooch had originally recommended the purchase of in 1854. Gyfeillon, now known as the Great Western Colliery, conflicted with their colliery interests and John Calvert, the original owner of Gyfeillon and former TVR contractor who had been recommended to Brunel by Stephenson, bought it back in 1864.[33] Gooch decided it was time to go and on 4 April 1864 he tendered his resignation.[34]

He was now free to concentrate on what would be the most constructive period of the working life of the *Great Eastern*, the laying of the Atlantic telegraph. With Brassey, Barber and others, Gooch formed the Great Eastern Steamship Co. to buy the ship at auction, provided she went for £80,000 or less. The bondholders were invited to join the new company and were persuaded to agree to the ship being sold at auction again, but this time with no reserve. Barber and Yates, the latter becoming the secretary to the new company, went to Liverpool to attend the sale, and to everyone's surprise the ship was knocked down to them. As Gooch would write; '…a ship that had cost a million of money and was worth £100,000 for the materials in her, was sold to us for £25,000.' The bondholders had the option of coming in to form a new company by taking shares equal to the amount of their bonds as fully paid up. To manufacture and lay the Atlantic cable a merger was formed of the Gutta Percha Co., who would make the copper core and the insulation of the Atlantic cable, with Glass Elliot who would complete the protective armouring of the cable and lay it. The new company was to be known as the Telegraph Construction and Maintenance Co. or Telcon for short. Gooch was one of the moving forces and sat on the board of the company. John, later Sir John, Pender MP (1815–96), a director in the business Gooch had formed to promote

Joseph Whitworth's gun developments, was to be chairman, and the role of managing director was taken by Richard, later Sir Richard, Glass (1820–73) of Glass Elliot. The Great Eastern Steamship Co., with Gooch as chairman, then chartered the *Great Eastern* for £50,000 of cable shares to Telcon who on 5 May 1864 formally took over the contract to lay the cable for the Atlantic Telegraph Co.[35] Telcon's price was £300,000 in cash and £237,140 in shares, and they would meet all the expenses of modifying and fitting out the great ship for what was to be her greatest role.[36] The year 1864 saw the launch of another Scott Russell ship, this time in Cardiff; the ship *Mallorca* being built by John Scott Russell's son Norman at the Bute Shipyard on the Taff.

The *Great Eastern* was not the only association with Gooch and underwater cables, his former chief draughtsman, Thomas Russell Crampton (1816–88), had gone on to lay the first successful cross-Channel cable in 1851. The *Leviathan* itself had caught the attention of the American telegraph entrepreneur, Cyrus Field, who visited the shipyard in 1857 to be told by Brunel that his ship was the ideal vessel to lay an Atlantic telegraph cable. With the failure of the second attempt at cable laying in 1858, it only lasted three weeks before failing completely, seven years would elapse before the third attempt was made. This time the *Great Eastern* was available, but even with the largest ship in the world success was not guaranteed and the cable was lost 300 miles from Newfoundland. 1865 was also a busy time for Gooch as he went into Parliament as the Conservative member for Cricklade and in November of that year he was invited back to the GWR – this time as chairman of the company. From the time of the loss of the cable to early the following year, 1866, Cyrus Field attempted to raise funds for a second attempt and Gooch records that he arrived at his house late one Sunday night in despair as all his means had failed. Gooch proposed to him that a new company be set up to which he and his fellow directors of Telcon would subscribe, and the

Following the end of her cable-laying days the Great Eastern *arrived at Milford Haven in 1876 where she made her way around the dock work taking place to a specially prepared gridiron a short distance from, but parallel to, the quay on the Milford town side, her great length being marked out on Hamilton Terrace above. (Roger Worsley collection)*

Anglo-America Co. was formed. The production of a new cable was put in hand and the *Great Eastern* was fitted up with new machinery for hauling in the cable along with other equipment that they thought would be useful in the light of the previous year's experience. At the beginning of July 1866 the ship set sail for Ireland having taken onboard a new cable at Sheerness. This was successfully paid out from the *Great Eastern* and the cable that had been lost the previous year was grappled, spliced and connected with the result that two telegraph cables were now in operation across the Atlantic. After two years of laying cables the *Great Eastern* was refitted and chartered by the French Government to carry the anticipated large number of visitors to the Paris Exposition of 1867, but only 191 paid for the trip, amongst them Jules Verne.

In 1869 the *Great Eastern* was to lay its third Atlantic cable, this time from Brest to Newfoundland for the Societe du Cable Transatlantique Francais. On the way to Portland, Gooch was entertained by fellow directors of the GWR, and associated companies, at the Royal Hotel, Weymouth, many of whom would be shown around the *Great Eastern* by Gooch. He recalled that they were much pleased by this, and that bridges were mended that day between an old adversary: 'Mr Talbot stated that altho' he had been opposed to my election as chairman, yet he was now very pleased he had been overruled.'[37] Just over two months after the *Great Eastern* returned from this expedition she was being made ready for another cable-laying run that would this time take her on a route that finally suited her original design brief. She was to steam to Bombay to lay a cable from there to Suez as part of the cable for the British Indian Submarine Telegraph Co. (the chairman of the company being John Pender) and left Portland on 6 November 1869.[38] After a long journey around the Cape of Good Hope, the *Great Eastern* arrived at Bombay on 28 January 1870 and on 14 February she spliced on to the shore end of the cable and started paying out towards Aden. Some 1,750 miles

The Great Eastern *attempting to leave the Milford Haven dock in 1880. (Roger Worsley collection)*

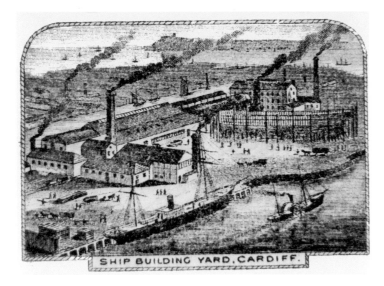

Bute Shipyard, the site of Scott Russell's yard on the Taff, as depicted in a Bute Shipbuilding, Engineering & Dry Dock Co. advertisement of 1891. (SKJ collection)

Left: *The old shipyard as it appeared from Hurman Street in 1910 when Curran Bros took over the site. (The Edward Curran Companies, A Review of Half a Century 1903–1953)*

Below: *The shipyard site in 1978. (SKJ photograph)*

THE NEW SPANISH MAIL STEAM-SHIP MALLORCA.

This page: Two illustrations of the ship built by Norman Scott Russell, the son of John Scott Russell, at the Bute Shipyard on the Taff. Said to have been later used as a blockade runner for the Confederate Navy, it was launched in November 1864, the ceremony being performed by Mrs Ann Clark, the wife of G.T. Clark. Ironwork for the first iron ocean-going ship built in Cardiff had been supplied by the Dowlais Works, of which Clark was the resident trustee. Working on the Mallorca *was Henry Keyes Jordan (SWIE president in 1883) who had followed Scott Russell from Millwall. (Lower illustration from* The Cardiff Times, *18 November 1864, both courtesy of Cardiff Central Library)*

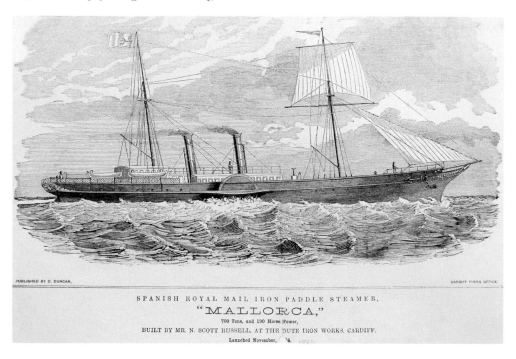

PUBLISHED BY D. DUNCAN.

CARDIFF TIMES OFFICE.

SPANISH ROYAL MAIL IRON PADDLE STEAMER,

"MALLORCA,"

700 Tons, and 190 Horse Power,

BUILT BY MR. N. SCOTT RUSSELL, AT THE BUTE IRON WORKS, CARDIFF.

Launched November, '64.

of cable were laid when she reached Aden on 26 February but rough weather prevented the final splice with the shore end at Aden until 2 March 1870. Gooch recorded that 'It is the first cable she had laid without my being on board.' In 1872 the *Great Eastern* laid her fourth, and last, Atlantic cable and two years later the *Faraday*, the first specially designed cable-laying ship, was launched. At the end of her cable-laying days the ship languished at Milford Haven, still the largest ship in the world but with no useful purpose.

CHAPTER 8 NOTES

1 *The Times*, 26 February 1853.

2 Neyland would be known as Milford Haven until 1859 when it reverted back to its proper name of Neyland for a few months until being renamed New Milford before finally reverting back in 1906.

3 Wilson, Roger Burdett, Ed. (1972), p.83, *Sir Daniel Gooch Memoirs and Diary* (David & Charles: Newton Abbot).

4 Wilson, Roger Burdett, Ed. (1972), p.83.

5 Both were iron paddle steamers, the *Scotia* being the last built for Cunard. See Stephen Fox, *The Ocean Railway* (Harper Perennial: London, 2003).

6 See Chapter 7; William Hawes was the brother of Benjamin Hawes, Brunel's brother-in-law.

7 *The Times*, 26 February 1853.

8 *The Cambrian* reported in its 17 August 1860 edition that the *Great Eastern* would arrive at Milford Haven on 25 August, but that was after the publication of the following week's edition so the story was covered in the 31 August issue.

9 Dugan, James, in his book *The Great Iron Ship* (1953) states that only a stand-by crew of twelve men under Alexander Maclennan were left. However, according to the *Steam Shipping Chronicle* for 12 October 1860, quoted in Emmerson, George S. (1977), p.98, *The Greatest Iron Ship: S.S. Great Eastern* (David & Charles: Newton Abbot). 'Captain Hall, Manager Bold and chief engineer McLennan and all but twelve of the crew had been dismissed.'

10 *The Times*, 19 September 1860.

11 *The Times*, 19 September 1860.

12 See Chapter 1, Henry Marc Brunel refers to him as having 'died of lungs'. HMB PLB 03/10/1865.

13 See photograph, she is supported for about 580ft of her whole length, on two grids of 150ft-long each, with an interval between of nearly 300ft of levelled beach.

14 Emmerson, George S. (1977), pp.97–98.

15 *The London Journal*, 21 March 1857.

16 Emmerson, George S. (1977), p.97. The white line had been added to the décor of the ship prior to its maiden voyage.

17 Beaver, Patrick (1969), p.70, *The Big Ship, Brunel's Great Eastern, A Pictorial History* (Hugh Evelyn Ltd: London).

18 Wilson, Roger Burdett, Ed. (1972), p.82.

19 *The Cambrian*, 29 March 1861.

20 *Glasgow Herald*, 27 March 1861.

21 Emmerson, George S. (1977), p.156, *John Scott Russell: A Great Victorian Engineer and Naval Architect* (John Murray: London).

22 Dugan, James (1953), p.86, *The Great Iron Ship* (Hamish Hamilton: London).

23 Wilson, Roger Burdett, Ed. (1972), p.84. Baker was a director of the GWR who had resigned in 1856 but returned in 1858. He was also a director of the ESN and had become chairman of the GSC.

24 Wilson, Roger Burdett, Ed. (1972), p.85. *The Cambrian* reported on the disastrous accident and that most passengers left the ship at Cork (20 and 27 September 1861 editions).

25 *The Cambrian*, 11 and 25 October 1861.

26 Leleux, Sydney A. (1965) p.24, *Brotherhood, Engineers*, David & Charles: NewtonAbbot. See Chapter 3 for further details of Brotherhood's connection with Brunel.

27 *Birmingham Daily Post*, 22 February 1862 (reporting from the *Haverfordwest & Milford Haven Telegraph*).

28 Wilson, Roger Burdett, Ed. (1972), p.84. The incident with the boat and the drowning of the two sightseers is not mentioned.

29 Dugan, James (1953), p.115.

30 *The Cambrian* talks about a rent 8oft long and 3ft wide, 24 October 1862.

31 Preece, William Henry (1860-1861), *On the Maintenance and Durability of Submarine Cables in Shallow Water*, Minutes of the Proceedings of the Institution of Civil Engineers, Vol.XX, Session 1860-1861, Discussion on Paper, p.61.

32 Emmerson, George S. (1977), p.117. The subscriber capital of the company was £403,404.

33 See Vol.I., p.213.

34 Gooch Platt, Alan (1987), pp.132–133, *The Life and Times of Daniel Gooch* (Alan Sutton: Gloucester).

35 Pender had been elected to the board of the Atlantic Telegraph Co.

36 This episode of her career is also well documented and beyond the scope of this book to record in detail but the emphasis on the Welsh connections with the ship will continue. The sources already quoted on the *Great Eastern* cover this period along with books such as the well-known contemporary account *The Atlantic Telegraph* by W.H. Russell and more recently Gillian Cookson's *The Cable: The Wire That Changed The World* (Tempus Publishing Ltd: Stroud, 2003).

37 Wilson, Roger Burdett, Ed. (1972), p.153.

38 Baglehole, K.C. (1969, third impression 1978) p.2, *A Century of Service: A Brief History of Cable and Wireless Ltd., 1868–1968* (Cable and Wireless Ltd: London). With the smaller vessel *Chiltern* they carried 2,600 miles of cable between them. John Pender was the chairman of the British Indian Submarine Telegraph Co.

9

DUKE STREET, DOCKS AND THE DEMISE OF DYNASTIES

'DES DOCKS ET DU CHARBON, DU CHARBON ET DES DOCKS...'[1]

Brunel's death marked the end of his Duke Street practice. However, it was not the end of No.18 Duke Street as an engineering office as Robert Pearson Brereton, his senior assistant, would take over the practice and finish the works begun under Brunel but were not completed at the time of his death.[2] In terms of a successor in the family, Brunel's elder son, Isambard Brunel (1837–1902), chose not engineering but ecclesiastical law, and it was his second son, Henry Marc Brunel (1842–1903), who was to follow his father and become an engineer. Under happier circumstances Henry Marc may have taken over the practice from his father, but he was just seventeen when his father died and there were financial matters to be resolved. There would be great changes to the office; William Barber (1817-76) sent out letters, the first, to sixteen of Brunel's engineers the day following his death.

This chapter will attempt to outline the engineering works in South Wales connected with Henry Marc Brunel, an engineer involved in his own right, and in partnership with John

Robert Pearson Brereton (1819–94), the man who completed the majority of Brunel's works left unfinished at his death. He started in Brunel's office in 1835 and remained until the end when he effectively took over the engineering practice. There was some criticism from the Brunel family, particularly Henry Marc, who felt he took undue credit on some of the works. Brereton remained at Duke Street, the Brunel family home and later offices for Henry Marc's partner John Wolfe Barry. It is believed the eye patch was a result of an explosion while working on the GWR in which he lost an eye. His signature is above, as written on the protrait. (Courtesy of the Brunel Society)

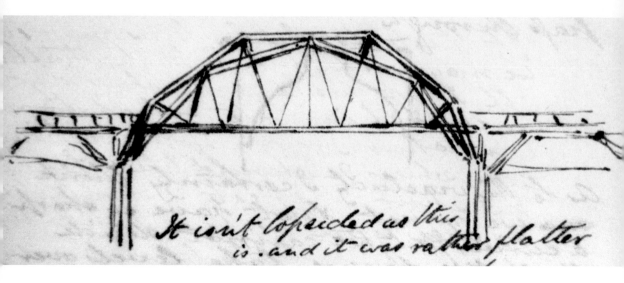

It isn't lopsided as this is. and it was rather flatter

Above: *Henry Marc's sketch of his father's Landore Viaduct in a letter dated 10 March 1861, with the writing underneath: 'It isn't as lopsided as this is, and it is rather flatter'. He was writing to Eddy Froude (1846–1924), William Froude's son, telling him that he stopped at Swansea on the way to see the* Great Eastern *at Neyland. Henry Marc and Eddy were planning to build a model railway, a broad-gauge model railway, of course, adding that an arch like Landore would '…look rather jovial in the middle of our viaduct…' (Courtesy of University of Bristol special collections)*

Left: *Henry Marc Brunel (1842–1903). (Courtesy of the University of Bristol special collections)*

Wolfe Barry (1836–1918), for many of the major works of the period, not just in South Wales but throughout Britain. One work that had involved him through many stages, from proposals and false starts through to final completion and from working on his own account to partnership with Wolfe Barry, was Barry Docks, the biggest dock works of the Victorian era. Henry Marc, somewhat understandably because of his father's stature, was destined to remain in his father's shadow, although in this later era of engineers he was described, in comparison to his partner Wolfe Barry, '… as a giant of different character'. A former pupil who had worked under Henry Marc and Wolfe Barry, Alexander, later Sir Alexander, Gibb (1872–1958), gave this opinion of the engineer:

> Of Henry Marc's own works the best known – London's Tower Bridge – reflects something of his original mind. He was never afraid to depart from accepted practice. Like others who choose to think for themselves, he has been called eccentric – and he probably was. We know at least that his whimsical humour often came to the youngsters as a welcome offset to Barry's steady drive.[3]

With Brunel's death occurring at such a time in his son's life, what were his educational and professional prospects? To anyone inquiring about how best to give their son an engineering education, Brunel would, according to Henry Marc, advise an ordinary education up to age sixteen, with special attention to mathematics. In Henry Marc's case this was done at Harrow and then King's College, London (which had established its engineering course in 1838), followed by a two-year period with his father. He would later ask Brereton to confirm in a statement the time he spent as a pupil of his father, from 11 May 1855 to 11 May 1857.[4] There then followed a 'gap' period in his formal education concerned with supporting his father in the run up to, and including, the launch of the *Great Eastern*, being near the family during his father's increasing illness and death and then supporting his mother. After his father's death he received support from one of his father's engineering friends and colleague, William Froude (1810–79). Froude had worked under Brunel on the Bristol & Exeter Railway, but left railway work in order to look after his widowed father, the Archdeacon of Totnes, in Devon.[5] He did not give up engineering, however, being involved in many local projects concerning roads, sea defences, agricultural improvements, the design of yachts and the improvement of Torquay's water supply, and also returned for a period to work on the North Devon Railway. Froude had committed a major error in his work on the North Devon for Brunel, but his contribution was valued despite him '…not perceiving the danger sooner…', and Brunel accepted it as being a 'very singular accident' and would not allow him, as he had offered, to forgo his salary.[6] Brunel's generosity of spirit on this and other occasions may have engendered the kindness Froude was to show to Henry Marc following Brunel's death.

At the age of nineteen Henry Marc was apprenticed to W.G. Armstrong as a premium apprentice for two years beginning in October 1861. Other notable apprentices included Charles Parsons, of steam turbine fame, and Hamilton Rendel (b. 1843), the son of James Meadows Rendel (1799–1856). Henry Marc's connection with Armstrong may have led to a further family link in 1864 when Brunel's eldest son, Isambard, married Georgina Noble, the sister of Armstrong's managing director, Captain, later Sir, Andrew Noble. Brunel's only daughter Florence (1847?–76) married a master of Eton College, Arthur James, whose daughter, Celia, also married into the Noble family, in this instance Sir William Armstrong Noble 3rd Bt. As both Henry Marc and Isambard would die childless, it is through this connection that the descent of the Brunel lineage was ensured.[7]

The works were established in 1847 by Sir William George Armstrong, later Baron Armstrong of Cragside (1810–1900). During this time at 'bloody Newcastle', as he referred to it in one of his letters, he would keep in contact with Froude over, amongst other thing, the design of a yacht called *Raven*. The likely cost of the premium given (if one was paid) to Armstrong was addressed by Henry Marc himself when writing to his mother some four years later in answer to a question raised by Mrs Holland on the cost of 'educating a civil engineer':'Civil engineering is a profession requiring special knowledge customarily acquired by working as pupil under an engineer in practice to whom a premium seldom under £250 or £200 is given.'[8]

It was fortunate that Brunel had such a good relationship with Armstrong, which no doubt lead to his son's pupilage on such a privileged basis. Certainly an engineering apprenticeship at Elswick would be a useful experience and one where he would gain an understanding of the new dynamics of hydraulic power, not that Elswick was confined to the manufacture of hydraulic equipment as it undertook general engineering work of all descriptions. From the Crimean War, Armstrong was developing new artillery, including breech-loading rifled guns.[9] He was also investigating the application of wire-wound barrels, which had interested him since 1855. Brunel had planned to commission Armstrong to make this gun but an existing patent held them back.[10] Henry Marc's letter books contain numerous references to the Armstrong gun and the rifle trials between Armstrong and Whitworth. HMS *Warrior* launched in 1860 had 110-pounder Armstrong guns; it was the first ocean-going iron-hulled battleship in the world and the most terrifying battleship of its day.

Before the end of this, of what Henry Marc refers to as a pupilage, he talks to Charles Saunders, the first secretary of the GWR, about approaching John, later Sir John, Hawkshaw in January 1863 for his view on settling his pupilage at Easter or Midsummer of 1863, and to ask his advice '…as head of profession.'[11] Saunders approved of Henry Marc's choice of Hawkshaw as the 'best man'. John Fowler was also considered but Saunders did not know what terms he

HMS Warrior, *built in 1860. Henry Marc's letter books contain numerous references to the ship which was Britain's first iron-clad battleship. Long after it had been decommissioned as a warship, it served as a floating oil jetty at Milford Haven. Seen here in the 1970s before it was taken to Hartlepool to be restored back to its former glory, on display at Portsmouth. (SKJ photograph)*

was on with Scott Russell and in the light of his father's experiences he would not work with any engineer who might be sympathetic to the Scott Russell 'camp'. Saunders did not like Hawkshaw as an individual but he thought '…Sir J Hawkshaw had most practice and [was] the better man.'[12] In October 1863 Henry Marc continued his apprenticeship as a pupil with Hawkshaw in time to be employed on the dock works at Penarth. Henry Marc records that in the first few months he was engaged in the drawing office on various projects including '… the Charing Cross Railway works…'[13] Penarth Dock had much hydraulic machinery supplied by Armstrong's, for working capstans, opening gates, taking ballast and small coal out of ships and for a swing bridge.[14]

The story of the Penarth dock has been touched on before as part of the original aspirations of the Taff Vale Railway to establish its own shipping facilities independent of Bute.[15] The revival of a dock at Penarth began with a meeting in February 1854 chaired by the Cardiff businessman John Batchelor (1820–83), at that time the Mayor of Cardiff (and Cardiff Police Magistrate). His business, a timber and ship yard run by him and his brother James Sydney Batchelor, had suffered due to Brunel's diversion of the Taff for the South Wales Railway.[16] The meeting included colliery owners, industrialists and shipping agents, all seeking improvements to the existing shipping facilities at Cardiff. On 10 July 1855 the Ely Tidal Harbour & Railway Co. was formed, which now sought parliamentary approval. Batchelor was to join the board of this company in November 1855 and an Act was granted in July 1856 to authorise the construction of the railway and harbour works. Hawkshaw was appointed consulting engineer and Messrs Rennie & Co. successfully tendered for the work in December 1856.[17] This, the first phase of development there, was too early for Henry Marc's involvement, the railway and Ely Harbour works being completed in 1858, the harbour opened on 18 July. A year earlier it had been decided to press ahead with the construction of a dock at Penarth and extend the railway to serve it. The company was to change its name to the Penarth Harbour Dock & Railway Co. to reflect the changes. The successful contractor this time was Messrs Smith & Knight, the contract being sealed in May 1859.[18] Hawkshaw was still involved, but this time with Samuel Dobson (1826–70) as joint engineer, and it was expected that the dock would be completed by the autumn of 1862. The dock consisted of the main dock which was 2,100ft long and 370ft wide, giving an area of 17½ acres and a basin which was 400ft long and 330ft wide (covering an area of 3 acres). There was a 270ft-long and 60ft-wide entrance channel from the basin to the dock, the sea gate being 60ft wide, and Penarth could be operated as a tidal harbour allowing ships to sail in and out at high water without locking. The depth of water was 35ft on spring tides and 25ft on neap tides, on the sill of the sea gates and the lock gates. This gave a depth of 4ft lower than the Bute East Dock and 7ft lower than the Bute West Dock sill. Nine stone piers were built on the southern side of the dock as the base of a single tip, except for the second pier which had two. A further two tips on the south side of the entrance basin completed the total of twelve tips. There were no coal tips on the north side of the dock; hydraulically operated cranes there unloaded iron ore and other cargoes.

Almost two years later the dock was not completed and hence the opportunity for Henry Marc to get involved. At first he was travelling back and forth, staying at Cardiff (Angel Hotel): '…off today to Cardiff for 10 days or a fortnight, in an awful hurry.'[19] He was to assist the draughtsman in producing a drawing of the dock as it was to be completed, being glad to learn something about stones and mortars and to get away from the manager at Hawkshaws, a Mr H. Hayter: '…awfully sick of the "head-man"'. He had clearly fallen out with Hayter, who he regarded as a fool, and going to Cardiff was one way of getting away from him. He writes from the hotel on 18 May 1864; 'Awfully hot here, a man died of sunstroke.'[20] Two days later he writes to his mother and states that the country is very pretty and the hotel is very

Close up of watercolour of Penarth Dock, Henry Marc's place of work, by G.H. Andrews, 1865. See colour section. (Courtesy of Glamorgan Record Office)

comfortable, the work site being about two or three miles from here and in '...a very pleasant situation'.[21] However, his real thoughts on Cardiff appear to be somewhat different, as writing to E.M. Froude on 22 May:

> Cardiff's not only a horrid place from its costive properties with regard to her acquaintances but has always looked upon it as a dirty place in itself as a dirty town with bad inns, his tradition based on once when IKB and IB were here the landlord of the inn occupied his leisure moments in dying of cholera as also did the judge of the circuit. However, has found the inn very good and the weather so splendid this week it's been quite jolly...[22]

A view of the dock looking towards the headland with the coaling appliances on the right. (Courtesy of National Museum and Galleries of Wales, WI&MM collection).

Close up of the Marine Buildings, alongside the Custom House at Penarth Dock. Although the TVR could not show their real involvement at Penarth because of the agreement entered into with Lord Bute, they could not resist having a florid 'TVR' woven into the wrought-iron balcony railings on the building! (SKJ photograph)

Henry Marc would return on 25 May to London, via Bristol, to work on various other projects for Hawkshaw. In the September of 1864 he records that he will return to the works at Penarth as there is little or no work in the office, estimating that there is £400,000 worth of work there and he will gain much experience of dock work. Hawkshaw was joint engineer of the work with Samuel Dobson with whom he got on very well: '…a very nice man.' Henry Marc would be Hawkshaw's only resident assistant on site: 'If S Dobson gives him anything to do shall like it very well as the works are about finishing and the bustle will be agreeable but really has learnt more engineering [here] … than for a long time at J Hawkshaw's.'[23]

Dobson was born in Northumberland the son of a farmer, and was apprenticed to a colliery viewer. In 1848 he was working for T.J. Taylor of Earlsdon in Northumberland who secured a position for him as mineral agent for the Windsor (later Windsor Clive Estate) and he afterwards went into business on his own account. He had branched into civil engineering and projected the Penarth Harbour Dock & Railway (PHDR) for which he and Hawkshaw were joint engineers. Elected a member of the Institution of Civil Engineers in 1856, he supported the formation of the South Wales Institute of Engineers a year later but was to die of consumption in 1870 at the age of forty-five.[24] It was some weeks before Henry Marc actually travelled down to South Wales and although writing from the Angel Hotel in Cardiff, he indicated that he had secured accommodation in Penarth itself by requesting that a copy of *The Times* to be sent to him c/o W. Richards, Bute Cottage, Penarth, the newspaper being one means of keeping him in touch with the wider world.[25] On the following day, 29 September

Bute Cottage, now known as Bute House, in Grove Place, Penarth, photographed on 16 May 2008. (SKJ photograph courtesy of Thomas and Pamela Evans)

Henry Marc's sketch of Bute Cottage, much like a house he would have drawn in his infant years, from a letter dated 14 October 1864. (Courtesy of University of Bristol special collections)

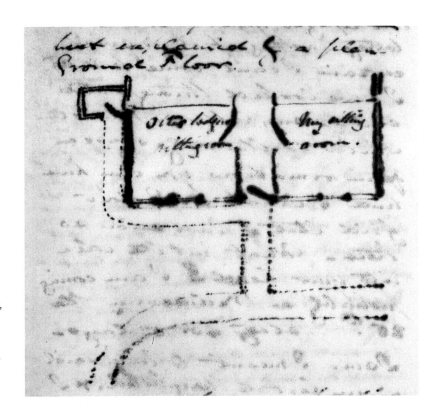

Another sketch of Bute Cottage, this time show-ing the ground floor. (Courtesy of University of Bristol special collections)

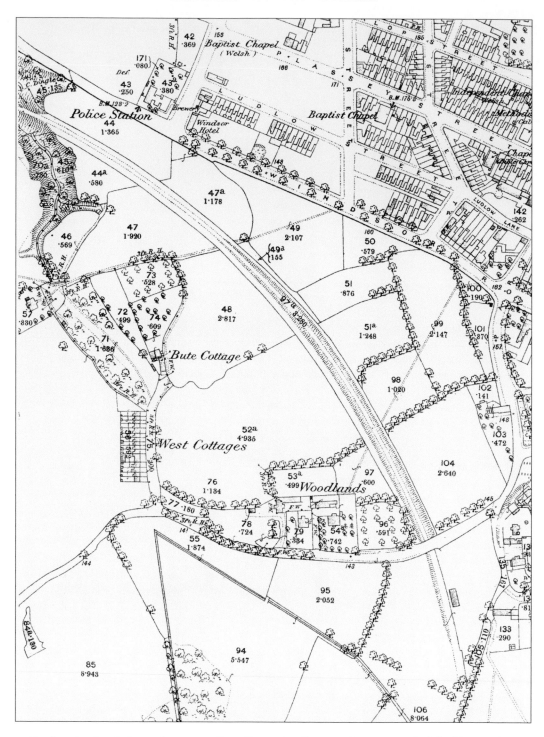

On the 25in 1879 Ordnance Survey map, Bute Cottage is still in a rural situation to the left of the recently built branch of the TVR. The first of the rows of terraced houses on the west side of the railway, which would eventually dominate that part of Penarth, has been built as 'West Cottages'. (Courtesy of Penarth Library, Crown Copyright reserved)

1864, he entered Bute Cottage for the first time and wrote to his mother asking her to send his waterproof leggings, waistcoats and books down: 'Has nicish lodgings the same as other J Hawkshaw's pupils who've been here.' Commenting that the weather was 'Splendidly fine' and that the position of the cottage, on high ground, looked healthy, he added that he '... came this morning'.[26]

On the tithe map Bute Cottage is shown in semi-rural isolation; at that time it was actually in the Cogan parish, and in the possession of the Marquess of Bute, hence the name 'Bute Cottage'. The house, gardens and fields which were leased to Thomas Evans, a local pilot[27] who, according to his descendent, Ernest T. Bevan, built the house.[28] In view of his comments above on the landlord of the Angel Hotel dying of cholera, was Brunel aware that both Thomas and his wife Mary died there of cholera, within a few days of each other, in 1854? The next occupant was William Richards, whose wife Mary is recorded as dying there in 1858.[29] On the 25in 1879 Ordnance Survey map, Bute Cottage is still in a rural situation with the Cadoxton branch of the TVR, originally the Penarth Extension Railway, running close by, but the first station in Penarth would not open for passenger traffic until 20 February 1878. Henry Marc called these new lodgings 'rather tooralooral'; an old detached house like a small farmhouse. It appears he was sharing the house with Richards, but there are no comments about him actually living at the house, although Henry Marc does complain about the space, particularly for his ablutions, as in order to shave he has to '...stand on one leg in his own light and overhanging a bath...'[30] Henry Marc went into more detail on the location of his lodgings a week after he first moved in, at the beginning of October 1864:

> A very pretty place, hasn't seen rain here. About 3½ miles from Cardiff. In a lodging, a small truly rural detached house among fields and picturesque honeysuckle sort of garden and orchard of sort of footway through fields to it sort of road immense way round at the back to get to it sort of place. A long way from a town, letters come and go once a day, not Sundays. 4/- to come out in a cab; got one of the people at the office to enquire if owner of the house could take him, he said he couldn't having other lodgers who had their own servants so HMB came himself with a confident air, took the place by storm and with the suavity of his manner got over the usual difficulty in lodgings getting the people to take the china shepherdesses, shell and feather bead baskets etc off the mantelpiece without hurting their feelings.[31]

At this time, around 3 or 4 October, he recorded that M. Rendel was down to inspect the works and Henry Marc found him at the docks, joined him on his inspection and went to dinner with him, which was 'very agreeable'. This would appear to be Alexander Meadows Rendel (1829–1918), the oldest son of the engineer James Meadows Rendel, although it is not recorded by Henry Marc if the works at Penarth impressed him in any particular way. There were no collaborations between Brunel and Rendel but some interesting crossovers; a few months before the opening of his chain ferry, Rendel's bridge design had been placed fourth in the Clifton bridge competition and his 1848 design for the Holyhead breakwater would include the last surviving section of broad-gauge railway.[32] Henry Marc would become acquainted with Rendel's other sons, Stuart and Hamilton, during their apprenticeship with W.G. Armstrong. Armstrong's supplied much of the engineering work including the hydraulic equipment for the docks and other engineering work such as the iron framework for the coal tips which were not hydraulic but self-acting. The hydraulic equipment would be fed from a wooden accumulator house and during very cold weather fires were lit near the tips to prevent the hydraulic system freezing.[33] The twelve coal drops could ship 150 tons of coal an hour on these self-acting or balance tips. The loaded coal wagons would be marshalled on the rail sidings leading to the tips where individual wagons would be detached from the train. They would run by force of gravity to the turntable and, if necessary, be turned so that the door end of the wagon faced the vessel, and it would then be run onto the cradle of the tip. The cradled truck would be lifted to the required level by hydraulic power, and then the back of the wagon would be held by a suspended chain and hook. Then, by lowering the cradle under the control of a brake, the wagon was tipped at a sufficient angle to meet the coal chute, and the coal was discharged into the ship's hold. After the wagon was emptied or 'tipped', weights which counterbalanced the weight of the cradle and the empty wagon would lift the cradle and empty wagon to the higher rail level where the wagon would be released to the return-ing 'empties' sidings. The tips dealt with wagons not exceeding a capacity of 6 tons, and were operated by three men and a boy.[34]

A plan he drew in a letter later in October was for his mother's benefit as he stressed that he has little privacy there and it would be unsuitable for his mother to stay.[35] Despite all his comments and complaints, Henry Marc appears to have enjoyed his stay at Bute Cottage as another sketch of the front elevation of the house takes him back to his childhood and early attempts at drawing:

> Bute cottage is chiefly remarkable for fulfilling the conditions of the house of HMB's infant years. What more can man desire? Has drawn it too tall for its width and left no room for a chimney. Hasen't set himself up with a woodpecker or a tapping or to his knowledge a beech-tree hollow or otherwise. A very pleasant place. Lots of ships and the Bristol Channel to look at. It's opposite Clevedon and the highest land hereabouts. Can see on a clear day as far as Ilfracombe.[36]

At Penarth, however, he did miss a 'good sherry' as when he wrote to his friend Metford invit-ing him to stay at the cottage, saying that if he arrived on a Sunday he would go to Cardiff church with him, but could he bring a good bottle of sherry as he had '…nothing but trash here.'[37] In the time between his first and last visit to Bute Cottage, and, indeed, throughout this time, Henry Marc filled his letters with news to family and friends as well as engineering matters which ranged from Hawkshaw's projects, Armstrong's work in general, the trials with Whitworth, the Dartmouth & Plymouth Railway, to the *Great Eastern*. On family matters he recorded his thoughts to his mother on the sale of Watcombe, which was to be his father's country estate: '…if the matter rested with him he should not sell the place for so small a sum…'[38] Watcombe was to have been the house from which Brunel would escape the 'madness' of the world, and which he had made the first purchase of land in 1847. Sadly the pressure of work meant that although he had carried out much landscaping of the grounds and planting of trees, by the time of his death only the foundations and cellar of the house had been built.[39] In November 1864, after returning from a break, he was back at work, which he liked very much, but was living in dread of being summoned back to town (Hawkshaw's London office). From Penarth he made occasional trips to Paignton and the family home at Duke Street. On 23 June 1865 he noted in a letter that he should stay there as long as he could hold on, he was doing the new scheme for Dobson and he acknowledged that this was learning and he mustn't leave it. If work stopped when the TVR took possession he couldn't stop much longer and would apply for a holiday. The following month he recorded a problem with the rollers on the dock gates, not with the rollers themselves but with obstructions on the rollers' path which gave trouble in opening and shutting them.[40] Delays with the dock gates had been the main cause of delays at Penarth; they were being manufactured by Cochrane's of Dudley, who had promised delivery in time for the proposed opening on 1 July 1864. Dobson visited their works at the beginning of May 1864 and was unhappy with the progress being made, progress that '…if not accelerated the

Aerial photograph taken in 1943 showing the dock and Ely Harbour. Penarth also suffered from the downturn in coal exports but during the Second World War was used for military purposes (note the barrage balloon). (Courtesy of Welsh Assembly Government Photographs Section)

Another view a year earlier showing the headland. (Courtesy of Welsh Assembly Government Photographs Section)

gates will not be in time to admit of the Dock being opened for traffic this year.'[41] Delays continued; at one stage Cochrane's blamed the delays on the masonry of the dock not being ready to allow the erection of the gates. Dobson refuted this and after much chasing of Cochrane's the dock was ready to be opened on 10 June 1865. Cochrane's, or more correctly Cochrane Grove & Co. of Dudley, had also been busy in 1864 on one of Hawkshaw's projects. This was the removal of the Hungerford Suspension Bridge and the transfer of much of the ironwork to Clifton. They also built the bridge that replaced Hungerford, the Charing Cross Railway Bridge, work on which occupied Henry Marc at the start of his time with Hawkshaw.[42]

His time at Penarth was coming to an end, and a holiday at Greenock was to take him away from Bute Cottage for almost a fortnight, and he would write to William Richards on 12 October 1865 to inform him that he was sorry he was leaving, he would be back on 18 October but leave for good the following day, and asked if he could forward his letters.[43] The dreaded summons had finally come from Hayter who requested his return to Great George Street (Hawkshaw's office in London), so he decided to call on Dobson before he left. Dobson wanted him to stay on and complete the entrance jetty to Penarth Docks, some six weeks worth of work, which he informed Hayter of. After returning to Duke Street he repeated that Dobson wished him to finish work there and that it would take a couple of months. In a letter to his mother, however, other Hawkshaw work would now claim his attention. In the same letter, after telling her about Greenock, he stated that he left his work in proper train: 'I don't think I ever passed so pleasant and happy a time as the greater part of my stay there.'[44] In a few months Henry Marc was engaged on Hawkshaw's Channel Tunnel project alongside E. Blount, and swapped Bute Cottage, Penarth, for the Grand Hotel in Paris.[45]

In all, Henry Marc was at Bute Cottage from 29 September 1864 to 19 October 1865, some thirteen months, believing the work would be completed in four or five months on a project that was almost three years overdue as far as the completion of the dock was concerned. Henry Marc was to work here after the formal opening of the dock itself, of which the decision had been taken to open the dock formally at high water, i.e. 7.00 a.m. on the morning of 10 June 1865. On that sunny Saturday a great crowd of several hundred had assembled; the official ceremony saw the presence of a number of TVR directors, including W.D. Bushell, a TVR director who had been involved in the TVR/Bute negotiations of the 1840s, William Sheward Cartwright of Llandaff, a prominent coal merchant, and James Poole (1797–1872), a Bristol coal merchant who was then the chairman of the TVR. Crawshay Bailey, the chairman of the Penarth Dock Co., was not present due to the death of his son-in-law. The joint engineers Hawkshaw and Dobson were also present but Baroness Windsor, who was to perform the ceremony, was not. She had got delayed en route and by 7.30 a.m. Poole took on the role, appropriately stating in his opening remarks; '…time and tide wait for no man, I have been requested to open the dock in her name. I do so and may God bless the undertaking. I now declare the Penarth Dock open.'

With that the gates swung open, and the first ship to pass through was the *William Cory*, a gaily decorated screw steamer which passed into the dock to the sound of cannon. The wife of Thomas Powell, another prominent coal freighter, christened her as the ship passed through. The steamer was followed by the lifeboat *Harriet*, also gaily painted and decorated, to be christened by the Baroness Windsor, who had apparently arrived shortly after the opening ceremony. The third ship to pass through was the new vessel, *Lady Mary Windsor-Clive*, a boat built by Batchelor Bros of whom John Batchelor had chaired the first meeting that led to the events on that day in June 1865. A number of demonstrations took place with the passage of the last ship, the Seaman's Mission boat, signalling the end of the ceremony which was performed by Baroness Windsor and the Hon. Robert Clive, who closed the dock gates hydraulically using the operating levers. The invited guests then adjourned to a sumptuous breakfast which had been prepared in a marquee.[46] At the breakfast banquet Poole would relate the story of the TVR's original proposal for a tidal harbour at the mouth of the River Ely and the arrangements they were forced to make with Lord Bute:

> The TVR Company having thus abandoned the conversion of the mouth of the River Ely, other parties took up the question, and had not contented themselves with merely making a tidal harbour, but they stand before you now as the proprietors of a great dock, nobly achieved, and scientifically and admirably carried out.[47]

In 1859 it had been decided not to complete the dock to full capacity because of the cost, but it was not until 1881 that work was undertaken to complete the work, increasing its acreage to twenty-six and a half. Hawkshaw was company engineer with H.O. Fisher as resident engineer. The contractor was T.A. Walker, the contractor for the Severn Tunnel, brought in by Hawkshaw, and the extension was formally opened in 1884 by Lord Windsor, on a date that was Brunel's birthday (9 April). This was not the end of Henry Marc's associations with South Wales; he refers to making plans for a dock at Cardiff in 1866 for the Glamorganshire Canal Co., but no Bill was deposited for this dock which would be sited on the east bank of the River Taff at its estuary.[48] Ten years later he would become involved in the story of Barry Dock.

Meanwhile, Brereton was busy completing Brunel's unfinished projects as well as some new projects of his own, which included Porthcawl. Until the beginning of the nineteenth century Porthcawl was no more than a small sheltered inlet, the main communities in the area being centred on the villages of nearby Nottage and Newton. The name Porth Cawl, or sometimes Porth-y-cawl, comes from the Welsh 'Porth' for port or gateway and the second part, 'Cawl', appears to refer to the appearance of the sea at this point where the prevailing wind churns the sea into a frothy mass ('cawl' meaning broth).[49] The idea of a harbour in this area was first mooted in 1818 when a group representing local interests met to discuss the best shipping place for coal and other minerals being exploited in the hinterland. Three options were looked at: the mouth of the River Ogmore, expanding the existing 'port' at Newton or the lee of a small promontory called Porth Cawl Point or Pwll (Pool) Cawl Point. The latter was chosen for its suitability for connection by tramroad which would have been more expensive to build from Newton whilst the Ogmore site was felt to be too unprotected for '…a safe and sufficient harbour'.[50] An application to present a Bill for a tramroad was reported by *The Cambrian* on 13 November 1824 as a line running from Llangonoyd, more correctly Llangynwyd, in the Llynvi Valley, to the parish of Newton Nottage and the new harbour of Porthcawl. The harbour and horse-drawn tramroad provided an important outlet to the sea for the developing iron and coal industries. Indeed, amongst the names of local landowners that were supported, the formation of the company included the industrialists Guest, Crawshay, Coffin and Buckland. A brief account of the Duffryn Llynvi & Porthcawl Railway and the broad-gauge Llynvi Valley Railway that followed it has been told in Volume II.

Brereton was also involved with the *Great Eastern* following the death of Brunel. Returning to the final years of the ship brings us to the mid-1880s. During this time ideas for using the ship came and went, all unrealised until on 28 October 1885 the ship was put up for auction at the Royal Exchange in London. Edward de Mattos was the successful bidder, acquiring for £26,200 the '…celebrated and magnificent Iron Paddle and Screw Steamship *Great Eastern*', as the auction particulars described it.[51] De Mattos was a director of London Traders Ltd and he proposed to use the ship to transport Welsh coal to Gibraltar where the ship would then

Porthcawl Dock in 1959; only the outer dock remains, but the shape of the filled-in inner dock behind, along with railway tracks, can still be seen. Designed by Brereton, the dock was opened in July 1867 and served by Brunel's broad-gauge Llynfi Valley Railway (LVR). Coal exports had virtually ceased by 1906 and the inner dock was abandoned but was not completely filled in until after the Second World War and Portcawl went from dock to seaside town. (Courtesy of Welsh Assembly Government Photographs Section)

be employed as a coaling hulk.[52] This was, ironically, the fate that would befall the *Great Britain* less than a year later when she was forced to seek refuge in the Falkland Islands during her last voyage. The proposal, however, ran into difficulties and following an approach from David Lewis's, the Liverpool department store, De Mattos's plans now changed. The managing director, Louis Cohen, wanted to promote David Lewis's, the country's largest drapers and universal suppliers, by using her as a floating exhibition in the Mersey to tie in with the 1886 Liverpool Exhibition of Navigation, Travelling, Commerce and Manufactures. He offered to charter the ship and set about engaging a master and a chief engineer to get the ship moving under her own power, choosing Captain Comyn and Peter Jackson respectively.

Getting the ship ready proved to be a most difficult task. The accumulation of debris around the anchors took four days to remove and 300 tons of marine growth had to be scraped from her hull.[53] Scott Russell's paddle engines proved to be beyond repair, so Jackson concentrated on the screw engines which were finally coaxed into life. Three times the engines were started but the frictional resistance and vacuum loss was too great to keep them going. Eventually the engines started astern and with violent protests kept growing at eight revolutions a minute, but when Captain Comyn signalled to stop and go slow ahead the ship stopped quickly enough but every effort to make her go forward was to no avail. The tide that day was lost and Jackson desperately needed an engineer who knew the engines intimately. Fortunately such a man was close at hand at Swansea in the shape of George Beckwith, now working there as a consultant engineer.

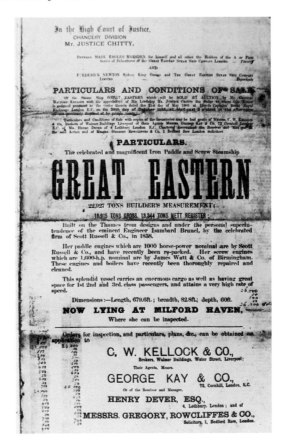

Auction notice for the Great Eastern, *'Now lying at Milford Haven'. On the right of this page can be seen the bidding figures which opened at £10,000 and increased by £50 bids to the selling price of £26,200. (Courtesy of Patrick Beaver)*

On receiving a wire sent by Jackson, Beckwith agreed to work on his old charge and arrived the following morning. Meanwhile, by working around the clock, air leaks were repaired and cocks, glands and valves were caulked.[54] By noon the following day Beckwith coaxed the engines into going full ahead. With the paddle floats removed and the wheels made fast to create as little resistance as possible in the water, she went at a rate of 5 knots, and with Captain Comyn at the helm, she left Milford Haven for the last time on 29 April 1885 steaming the 194 miles to Liverpool where she arrived on 1 May.

Signwriters onboard were busy during the voyage, turning the ship into a giant floating advertisement, dangling over the sides and painting legends such as 'LEWIS'S ARE THE FRIENDS OF THE PEOPLE'. The Liverpool Exhibition was to open on 7 May and this time was used to prepare the ship for its role as a showboat. The main and after cable tanks served as a huge music hall and the Grand Saloon would cater for thirsty exhibition goers. A dining room was now the function of the ladies' saloon and cheap furniture and furnishings obscured the last vestiges of the luxurious ship that was the *Great Eastern*. At a shilling a head, including the cost of the ferry to the ship, visitors could wander around the ship, taking in the music hall, bar and restaurant, along with stalls and sideshows on deck. Orchestral music concerts took place, there was dancing every afternoon with sacred music scheduled for Sundays, the latter no doubt to placate those opposed to the ship being open on the Sabbath. Lewis's use of the *Great Eastern* was a commercial success, and with the charter completed de Mattos was free to carry out his original plan. However, Gibraltar was no longer interested in the ship and de Mattos decided to try and replicate the success of the ship being used as a showboat. She was towed to Dublin and then back to Liverpool to repeat her role. His schemes failed to pay although he tried a third time towing the ship this time north where she was moored on the Clyde near Greenock. The *Great Eastern* was towed on all these voyages as it was felt too problematical to get the engines going again. The curse of the *Great Eastern* could be seen to be at work in terms of de Mattos' enterprises, and he attempted to auction off the showboat. On 20 October 1887 de Mattos attempted to sell the *Great Eastern*, then lying in the Clyde, at public auction, the fifth auction the ship had been through, but de Mattos had put in a bid of £26,000 through his manager to prevent the ship going for too low a sum.[55]

Some time later an offer of £16,000 was received from Henry Bath & Sons of Liverpool, London and Swansea, and de Mattos decided, very quickly, that this was the best he was going to get and sold the ship before they changed their minds. However, there seems to have been some delay before this became public knowledge, presumably because Henry Bath & Sons were exploring other options for the ship, one of which included refitting the ship for new uses. Henry Bath & Son remain in business today as the Henry Bath & Son Ltd group of companies, but the Swansea connection came to an end in the 1920s due to the Depression.[56] In 1822 Henry Bath was in business at Bath's Copper Ore Yard in Bath Lane, Swansea, in a partnership with his son, Henry Bath II. In 1846 Henry Bath II established the Swansea Iron Shipbuilding Co. and in 1849 launched the steam yacht *Firefly*, intended for use on the coast of Chile, and the iron steamer *Augusta* which was engined at Llanelli by W.H. Nevill. His ships developed trade with Chile, exporting coal and returning with copper ore and sodium nitrates. At one time as many as thirty ships made up the Henry Bath & Son fleet, with many of the 'Cape Horners' being built at Bideford in North Devon. The fleet and the copper yards were managed by Henry Bath II and his brother Charles who opened a London office in the 1850s. This was the start of a wider trading in metals, and in 1875 the company opened an office and warehouses at Liverpool.[57] Ship breaking was a lucrative source of metals and a business that Henry Bath & Sons were in, hence their bid for the *Great Eastern* in 1888, coincidentally the year that the last of the founder's sons, Charles Bath, died.

In *The Times* for 25 June 1888 it was reported that 'Henry Ball (sic) and Sons of Liverpool' had come to an understanding with the Naval Construction and Armaments Co. (Ltd) providing that Captain Barnett RN, the harbourmaster at Barrow-in-Furness, would allow the *Great Eastern* to be taken into the Ramsden Dock there. This could be done if the paddle wheels were removed, thus reducing her width from 120ft to 83ft, and she went in on the top of the spring tides. The Ramsden Dock had just been completed and it was just long enough to accommodate the hull and the dockgate was 100ft wide. The report in *The Times* suggested that it was for the purpose of breaking up the vessel or in altering her in some way (i.e., fitting her up with new machinery and putting the ship to work in the cattle trade, or employing her as a bulk petroleum carrier). Dugan, the author of *The Great Iron Ship*, tells the story that the representatives of Henry Bath & Sons, Messrs Morrice and Greer, began to get carried away by the dream of restoring the ship to her former glory, and Captain Barnett, no doubt worried that the ship might outstay her welcome as she had done in Milford Haven, turned down the proposition.[58]

With that option closed to them, Henry Bath & Sons decided to bring the ship down from Greenock to Liverpool, which was undertaken by engaging a casual crew of 100 'runners', and, escorted by the steamer *Steamcock*, she left the Clyde on 22 August 1888. It took three days coaxing 4 knots out of the screw engines and taking a tow from the *Steamcock* to reach the Mersey. There was some degree of fear amongst the runners when she encountered a storm, rolling heavily and losing her tow, but after four hours adrift she regained her tow and eventually resumed her place on the New Ferry gridiron at New Brighton. The journey and the uncertainty of her fate at Liverpool was noted by Gooch in his diary for 26 August 1888: '…I would much rather the old ship was broken up than turned to any base uses. I have spent many pleasant and many anxious hours in her, and she is now the finest ship afloat. Some good use might have been made of her.'[59]

The *Great Eastern* was still the largest ship in the world and the scale of work in scrapping the ship was five or six times as large as any ship breaking that had been attempted before. The work, it was reckoned, could be done by 200 men in a year at a cost of £20,000, and, contrary to normal practice, the parts of the ship would be auctioned off before she was dismantled. As well as the metals and the usual scrap materials of the ship, the fittings were much sought after by collectors of *Great Eastern* memorabilia. Following the auction, if things went to plan, Henry Bath & Son stood to make £58,000 less the purchase price and the costs of dismantling (i.e., £36,000). Work began on 1 January 1889 and the interior fittings were easily removed. The destruction of the double hull, however, was another matter. Cutting by oxyacetylene torch was in the future, and the men had to work the rivets loose by endless hammering, wedging, levering, chiselling and sawing. One of the firms involved at New Brighton was Messrs H. & C. Grayson, a long-established Merseyside shipbuilding and ship repairing firm dating back to 1760. The task of cutting up the ship was one of the tasks undertaken by Henry, later Sir Henry, Mulleneux Grayson (1865–1951) when he joined the family firm.[60] The innovation of the wrecking ball, using a steel ball swung from a derrick to shake the rivets loose, was one solution arrived at to speed up demolition. It was still a difficult and labour intensive operation and the noise generated must have been excruciating for anyone living within earshot. It was to take 200 men working around the clock two years to reduce the ship to scrap, and keeping true to the tradition of loss making that plagued the *Great Eastern*, all profits turned to loss. On 15 October 1889 Gooch passed away, but the work of the ship breakers was a long way from being finished and eighteen months passed before they got to the double bottom of the hull. Gooch's last comment on the ship, made in his diary following the auction on 21 November 1888, was: 'Poor old ship, you deserved a better fate.'[61]

CHAPTER 9 NOTES

1 Translating as 'Docks and Coal, Coal and Docks', a remark about Cardiff that it was 'nothing but', made by the artist Alfred Sisley who would stay in Penarth in 1897. I am grateful to Charlotte Topsfield of the National Museum of Wales for this information, in a letter from Sisley to the art critic Adolphe Tavernier, 9 July 1897. (Foundation Custodia, Institute Nearlandais, Paris)

2 There would be some new contracts, such as Porthcawl Docks.

3 Harrison, Godfrey (1950), p.30, *Alexander Gibb: The Story of an Engineer* (Geoffrey Bles: London).

4 Brunel Archives, Bristol University. HMB PLB 25/09/1864, letter to R.P. Brereton following an enquiry from Hugh Carlile.

5 Tudor, Geoff, 'A Brunel-Froude Engineering Dynasty?' (one of the 'might-have-beens of family history) in the *Genealogists' Magazine*, pp.23–26, Vol.9, No.1, March 2007.

6 Brindle, Steven (2005), p.233, *Brunel the Man who built the World* (Weidenfeld & Nicholson: London).

7 Buchanan, Angus (2002), p.195, *Brunel: The Life and Times of Isambard Kingdom Brunel* (Hambledon and London: London). Arthur James's impression of Brunel is given in Tudor, Geoffrey and Hilliard, Helen (2007), p.110, *Brunel's Hidden Kingdom* (Creative Media Publishing: Paignton).

8 HMB PLB 21/07/1865. Henry Marc also outlines entry to general business, as an engineer, contractor's assistants and mechanical engineering.

9 This was the Elswick Ordnance Works, set up because of Armstrong's Government positions, on Government committees and appointments for the War Department.

10 The patent had been taken out by Longridge, who never implemented the idea.

11 HMB PLB 24–25/01/1863.

12 HMB PLB 25/01/1863?

13 Brunel Archives, Bristol University. Letter by H.M. Brunel outlining professional career to date; 1 February 1877. H M Brunel Collection.

14 HMB PLB 05/10/1864

15 See Vol.I, Chapters 9 and 11, and Vol.II, pp.118 and 125.

16 Vol.II, p.116.

17 Thorne, Roy (1984), p.18, *A History of Penarth Docks* (County of South Glamorgan: Cardiff). Rennie was to employ James Hemingway, a young member of the famous Cardiff family, on the harbour works.

18 Thorne, Roy (1984), p.21.

19 HMB PLB 16/05/1864

20 HMB PLB 18/05/1864

21 HMB PLB 20/05/1864. He records travelling by the Bristol & South Wales Union Railway, the railway ferry his father designed but did not live to see complete. He met J.E. Benedict on the journey.

22 HMB PLB 22/05/1864. He comments that work is about three miles off, so getting into state of rude health.

23 HMB PLB 17/09/1864.

24 Minutes of the Proceedings of the Institution of Civil Engineers, 1870, pp.217–8. In 1869 he married Olga M. Julia De Smitt of St Petersburg.

25 HMB PLB 28/09/1864.

26 HMB PLB 29/09/1864.

27 Thorne, Alan (1997), p.33, *Place Names of Penarth* (published by the author: Penarth). Occupation given as pilot in the 1841 census but farmer in 1851.

28 Bevan, Ernest T. (1949), p.4, *Old Penarth*, typescript article in Penarth Library Local Studies Collection. Bevan believes the house was built a short time before their death, but Alan Thorne (above) suggests a date of around 1830.

29 William Richards's wife Mary was sixty-two at the time of her death. In 1865 Richards appears to have a new wife and son (born on 10 February 1865), the house is also known as Bute Villa. From *The Cambrian*.

30 HMB PLB 11/10/1864.

31 HMB PLB 02–03/10/1864.

32 Completely isolated from the rest of the broad-gauge network, it was built to convey the huge stone blocks used for the construction of the breakwater, see Vol.II, p.248.

33 HMB PLB 05/10/1864.

34 Thorne, Roy (1984), p.29, *A History of Penarth Dock* (County of South Glamorgan: Cardiff).

35 HMB PLB 13–19/10/1864.

36 HMB PLB 24/10/1864.

37 HMB PLB 10/11/1864.

38 HMB PLB 11/10/1864.

39 Tudor, Geoffrey and Hilliard, Helen (2007), p.151.

40 HMB PLB 13/07/1865

41 Thorne, Roy (1984), p.26. Cardiff. Quoted from Penarth Dock & Railway Co. minutes, 02/05/1864.

42 Hungerford was to supply 1,040 tons of ironwork for Clifton, and some 200 tons of new links were required for the additional third link on each side supplied by Cochrane Grove from the Lord Ward Round Oak works in Dudley. Vol.I, p.90.

43 HMB PLB 12/10/1865.

44 HMB PLB 20/10/1865. He writes to Richards (PLB 23/10/1865) a few days later regarding his unpaid wine bill!

45 HMB PLB 09/03/1866.

46 Thorne, Roy (1984), p.27.

47 *Cardiff and Merthyr Guardian*, 16 June 1865.

48 Prothero, Iorwerth W. (1994), *Barry Docks and Railways, Volume I* (published by the author: Barry).

49 Morgan, Alun (1987), p.35, Porthcawl Newton and Nottage (D. Brown & Sons Ltd: Cowbridge).

50 Morgan, Alun (1987), p.40.

51 Beaver, Patrick (1969), *The Big Ship* (Hugh Evelyn Ltd: London). Auction notice p.121.

52 Emmerson, George S. (1977), p.141.

53 Beaver, Patrick (1969), p.121.

54 Dugan, James (1953), p.204.

55 Dugan, James (1953), p.214.

56 The copper trading business was established by Henry Bath in 1794, a Cornishman born in Falmouth in 1776. In the 1920s the Swansea operation, along with the Bath family home at Alltyferin, was sold.

57 The office was in Bentinck Street, Liverpool. The very first London Metal Exchange warrant, issued on 20 December 1883, was completed by the company.

58 Dugan, James (1953), p.214. 'Captain Barnett saw the fever in their eyes and refused to let them have the dock.'

59 Wilson, Roger Burdett, Ed. (1972), pp.351–352.

60 *The Times*, 29 October 1951, obituary notice.

61 Wilson, Roger Burdett, Ed. (1972), p.352. Again Gooch had been following the story from the *London Standard*. C.R.M. Talbot, who had supported Brunel and was a director of the Eastern Steam Navigation Co., passed away on 17 January 1890.

10

BEYOND BRUNEL

'...ONE OF THE LAST BURSTS OF VOLCANIC ENERGY...' [1]

The above quotation, by the Rhondda-born author Gwyn Thomas (1913–81), is a reference to the extraordinary growth of the Port of Barry in the late nineteenth century, but could also be applied to the activities of the generation of engineers after Brunel. Barry was the largest docks undertaking of the time – not the first integrated railway and dock system, but it was the last of its kind.[2] There would be major docks built after this in South Wales, such as Cardiff's Queen Alexandra Dock of 1907 or the 1909 King's Dock at Swansea, but these were extensions to existing dock systems. Barry was a completely new dock complex, a new burst of energy, and it can be seen as the last in a line of industrial developments in South Wales, and one that was designed by the last of the line of an engineering dynasty, Henry Marc Brunel. In Barry such volcanic energy had long been spent, but the docks and its townscape remain an impressive monument of its late Victorian past. Where before there was little save rural villages and farmland, there was now a town that grew in parallel with the building of the docks, or, to quote Gwyn Thomas, a town that exploded into being: 'The town in its modern form blew into being as one of the last bursts of volcanic energy that brought Britain through the last century and up to the First World War.'[3]

The idea for building docks at Barry appears to have been first suggested in 1861 by a tenant farmer on Barry Island.[4] His farmland would cease to be isolated from the mainland when the docks were built, with a causeway and viaduct carrying railway and road being constructed to the island. Barry Island would become a popular Victorian seaside resort, a far cry from the smugglers that had operated there in the eighteenth century. The smuggling gangs gave way to sea-bathers and the island become popular for summer outings. In 1856 the iron master Francis Crawshay bought Barry Island and built the Marine Hotel '...for the more select Victorian visitor' in 1858, along with a landing pier for his yacht.[5]

In purchasing an estate on the coast, Crawshay was following the trend for industrialists to get themselves and their families out of the crowded environment surrounding their works and escape the diseases so prevalent in this industrial period. It also reflected on their social status. Lady Bell, the wife of the Middlesbrough ironmaster, observed that; '...the distance a man lived from his daily work was in direct proportion to his success in it.'[6] In 1836 John Guest purchased the 1,000-acre Sully Estate for £3,500, but Lady Charlotte did not care for it even though the romantic and secluded setting of the house, looking across to Sully Island, had inspired Charles Kingsley to take up Holy Orders in 1841.[7] In 1846 it was leased to the coal freighter Thomas Powell (1784–1864) the Guests purchasing the Wimborne Estate in Dorset that year. Powell was an important member of that union of ironmasters and coal freighters that came together in 1835, a second attempt at bringing the 'Merthyr Tydfil and Cardiff Railway' to fruition as the Taff Vale Railway.[8] Another member of that union was Walter Coffin, who had purchased Llandaff Court.

Sully was selected by Telford as part of his suggested improvements to the Seven Crossings. It was also where John Evans, the general manager of the Dowlais works, was to retire to in 1856, by which time the estate had passed to Guest's son, Sir Ivor Bertie Guest Bart, who was to become the first Lord Wimborne, and the development of the estate for housing became a factor in the promotion of railways to Barry.

In 1861 the one-time Barry Island farmer John Thomas, now in retirement at Merthyr Dyfan, a parish on the coast near Barry,[9] decided to write to J.W. Nichol-Carne of Dimlands Castle near Llantwit Major. Nichol-Carne was a director of the Llantrisant & Taff Vale Junction Railway, a railway backed by the TVR and incorporated on 7 June 1861 to convert the old Llantwit Fardre tramroad into a railway from the main line of the TVR near Treforest to a junction with the Ely Valley Railway at Maesaraul, one-and-a-half miles north of Llantrisant.[10] John Thomas had seen an article in the *Bridgend Chronicle* in the July of 1861 concerning the Cowbridge Railway and he decided to write to Nichol-Carne to '...offer a few remarks, in the form of a letter to yourself as a gentleman who possesses a great weight and influence in the County and also as a Director of the Llantrisant and Taff Vale Junction Railway...'[11]

In his letter he expresses a number of points, particularly that the scope of the proposed Cowbridge Railway is insufficient for the agricultural and commercial interests of the wider locality, in particular the parishes of St Marychurch, Flemingston, St Athan, Aberthaw, Penmark, Porthkerry, Sully, Cadoxton, Llancarvan and surrounding districts. He suggests that a branch railway, either on the broad or narrow gauge, be formed:

> ...to be called the Glamorgan 'Coast Railroad' adjoining the South Wales Railway at Pencoed as its western extremity and Penarth as its eastern extremity and taking the following direction, namely:-
>
> From Pencoed, by Llansannor to Cowbridge, 4 or 5 miles. From Cowbridge by St Marychurch, Flemingston, St Athan to Aberthaw, 4 or 5 miles. From Aberthaw by Penmark and Porthkerry to Barry Harbour, 6 or 7 miles. From Barry Harbour by Cadoxton and St Andrews to Penarth, 5 miles.[12]

The last two sentences contain a reference to the real value of such a railway – access to Barry Harbour, of which John Thomas goes on to say that the existing harbour could be improved in order '... to accommodate vessels of the largest size and heaviest tonnage.' In addition to limestone, coal and iron from the mineral districts could be shipped from Barry. In his reply Nichol-Carne suggested to John Thomas that he sent a copy of his letter to the Cardiff newspaper in order to circulate his suggestions more widely. This he did, and the letter appeared in the *Cardiff and Merthyr Guardian* on 3 August 1861. Even before John Thomas's letter, however, the suitability of Barry as a site for a major dock undertaking had crossed a number of promoter's minds from the Ogmore Valley Railway (OVR).

Amongst these was Henry Voss, the engineer and manager of the Ely Valley Railway (EVR) and one-time assistant engineer under Brunel and William George Owen, who in 1860 attempted to promote a standard-gauge line from Llantrisant to Cardiff.[13] The EVR was leased to the GWR the following year and, frustrated with his attempts to get the GWR to lay a third rail to allow narrow-gauge traffic to run to Cardiff and Penarth Docks, he teamed up with his brother-in-law, John Pyne, to pursue other alternatives. A new dock to the west of Cardiff was one idea, and by linking with the standard-gauge Ogmore Valley Railway (OVR) it might have been feasible. The OVR was backed by the firm of John Brogden & Sons, the major industrialists in the Llynvi and Ogmore valleys.[14] The Brogdens felt hemmed in by the

The first (No. 1) dock to be opened at Barry in 1889. Arguing the case for building a dock at Barry, David Davies stated, 'we have five million tons of coal, and we can fill a thundering good dock the first day we open it.' (SKJ collection)

'Sail and steam at Barry, all waiting for Rhondda coal.' (SKJ collection)

broad gauge and had promoted the standard-gauge OVR, which was to merge, despite the difference in gauge, with Brunel's Llynvi Valley Railway (LVR) in 1866 to form the Llynvi & Ogmore Railway (L&OR). Pyne was the land agent, surveyor and steward to Sir Ivor Bertie Guest Bart and, because of Pyne's position, Lady Charlotte discussed their proposals for branch lines as they affected the Guest estates.[15] Pyne and Voss approached the owner of the Wenvoe Estate, Captain Robert Francis Lascelles Jenner (1826–83),[16] with a suggestion for the construction of a dock at Barry as an alternative shipping place. Jenner was a major landowner in the Barry area and the estate had once included the leasehold of Barry Island. Voss was to have second thoughts about Barry and urged the EVR to promote a Bill to make the standard-gauge connection from its line at Gellirhaidd, a few miles north of the SWR's Llantrisant Station, to Blackmill in the Ogmore Valley, with powers to lay a third rail from the Llantrisant Station line to Gellirhaidd, but the Bill was rejected.

Jenner, however, decided to go ahead with the suggestion and, with the support of Sir Ivor Guest and others, promoted a Bill which became the Peterston to Cadoxton-juxta-Barry Railway Act of 1865.[17] It did not include a dock as Jenner's solicitor urged against it in view of the heavy expenses that would be incurred. The successful Act empowered the GWR to work the proposed line and, if desired, to lay a third rail outside the standard-gauge line to take broad-gauge rolling stock. It also included a branch to Sully, which the Guests wished to develop for residential property.[18] Three further Acts the following year included a tidal harbour, an extension of the 1865 line to the harbour quay and a branch off the Peterston to Barry at Cadoxton to Cogan.[19] The first was 'An Act for converting the Estuary of Barry Island in the County of Glamorgan into a Tidal Harbour...' This was to be a 600-yard-long quay and could be constructed cheaper than an enclosed dock. The arrival of engineers to carry out the survey work was noted in 1865: '...engineering staff arrived in the district for the purpose of making out the line of railway between Cogan and Barry for the late Captain Jenner of Wenvoe castle...'[20]

The Wenvoe Arms in Cadoxton was a popular hostelry for the engineers who left Barry in 1866 but who were to return the following year to complete their plans for the railway.[21] One of the engineers to stay at the Wenvoe Arms was a 'Mr Binney', more correctly Alexander, later Sir Alexander, Richardson Binnie (1839–1917), who later became the chief engineer for the London County Council (LCC) and would be president of the Institution of Civil Engineers in 1905. As chief engineer of the LCC he was to build the Blackwall Road Tunnel under the River Thames (in consultation with Sir Benjamin Baker and James Greathead) and the Greenwich Foot Tunnel.

Jenner was the first and only chairman of the Barry Railway Co., but there was little take up when the shares went on offer in 1866 because of the financial crisis that year with the collapse of Gurney, Overend & Co., and less than 10 per cent of the share issue of £160,000 was taken up.[22] Despite a number of attempts at reviving the Barry Railway scheme, as in 1870 when Richard Price Williams (1827–1916)[23] and one of the Barry Railway Co.'s engineers, J. Tolme, met unsuccessfully with the TVR Board to promote a railway from Bridgend to Barry, the company was wound up in 1874.

Barry remained, literally, a rural backwater, with small sailing vessels entering the old harbour to take on cargoes of limestone and to find refuge from storms in the Bristol Channel. The demands of the South Wales coal trade, however, would not leave Barry alone, particularly with growing complaints about shipping congestion at Cardiff. A scheme of 1876 backed by the Plymouth Estate (later the Windsor Estate) Trustees was intended to promote coastal property at Barry, and, although unsuccessful, would create much interest again – interest that would bring Henry Marc Brunel into the story of Barry Dock. At the end of November 1876 the Windsor Estate mineral agents, Messrs Brown & Adams, approached Henry Marc.

Adams had worked with Henry Marc on the Penarth Dock (Samuel Dobson had died in 1870), and they wanted him to take over their engineering commission for a dock at Barry and the revived Peterston to Barry line serving it. As the Windsor Estate had a vested interest in the prosperity of Penarth, Iorwerth Prothero believes that such a move was to avoid any charge being levelled against the Windsor Trustees of favouring Barry over Penarth.[24] Henry Marc responded quickly and in days had drawn up a plan for an enclosed dock of 70 acres with twenty tips or staithes. Funding the scheme remained the biggest hurdle, and the GWR agreed to construct the railway line serving it but only if the financing of the dock was assured. Here the scheme floundered, but two years later the proposals for a dock transferred as part of the lease of Barry Island, then in the hands of S.A. Tylke, to the Windsor Estate (who earlier that year had bought the freehold). Thus the Windsor Estate now had a vested interest in the development of Barry. In October 1880 concern over the reluctance of Lord Bute to increase dock capacity saw the setting up of a joint committee by Cardiff Corporation and Cardiff Chamber of Commerce, and in the same year, when Henry Marc reported to the Freighters' Association of Cardiff about the potential of five alternative sites for a dock outside Cardiff, it was Barry he recommended.

It is at this point that coal owner David Davies of Llandinam comes into the story of the Barry Railway in an official capacity (as he would have known about Henry Marc's plans in 1876). Davies usually gets the credit for the development of Barry, although that should probably be given to Archibald Hood and J.O. Riches, the former being a coal owner in the Rhondda Fawr and the latter a fellow director with Davies. Both Hood and Riches had sat on the sub-committee of the 1880 Cardiff Corporation and Cardiff Chamber of Commerce joint committee and had protested when Lord Bute was considering building a smaller dock than the one authorised by his 1874 Act; Riches stated that if they could not have facilities at Cardiff Docks they would go elsewhere.[25] The stage was set for another revival of Barry Dock, the promoters of which were to sign Heads of Agreement with local landowners in November 1882. These included Jenner, Lord Windsor and the Romilly Estate, who would receive £1 15s per acre per annum as rent and royalties of ½d per ton on minerals and building materials shipped or landed and 5 per cent of receipts from other traffic, including passengers. The Bill deposited the following year for the Barry Dock & Railways was to be passed by the Commons but rejected by the Lords. On 30 April 1884, a second Barry Dock Bill was presented and fiercely opposed, but on 14 August 1884 it gained the Royal Assent. Two months later on 14 November 1884 a sod turning ceremony took place at Castleland Point. Lord Windsor (later Earl of Plymouth) was to be the first chairman of the Barry Dock & Railway Co. and performed the task of turning the first sod, but the show was stolen by Davies who then showed him how to do it properly!

The engineering of the dock and railway was undertaken by the Brunel and Wolfe Barry partnership, with Sir John Wolfe Barry as engineer-in-chief, other colleagues including Thomas Forster Brown and G.F. Adams. Sir James W. Szlumper, who had been the engineer for all of David Davies's lines since 1861, was responsible for the railway section from Trehafod to Tonteg. John Robinson, who would write papers on the Barry Dock works and the Graving Docks for the Institution of Civil Engineers, was the resident engineer.[26] Engineering works included tunnels at Trehafod and Wenvoe, the latter, at one mile 107 yards long, being the sixth longest in Wales and the eighth longest on the British Rail Western Region. Later works on the Barry Railway's 'cut-off' routes included the Walnut Tree viaduct crossing the Taff Valley at Taffs Well and the Llanbradach and Penyrheol viaducts on the Rhymney Valley route. Two special trains carrying 2,000 invited guests passed through the tunnel on their way to the official opening of the dock on 18 July 1889. The major contractor was Thomas Walker, who was responsible for the dock itself and the line to Saint-yr-nyll, with Lovat and Shaw responsible from Saint-yr-nyll to Treforest and J. McKay & Son the section from Tonteg to Trehafod. Alexander Gibb, approaching the end of his pupilage with Brunel and Wolfe Barry, was set to work designing steel viaducts for the coaling berths and the layout of the complicated dock railway system.[27] A new dock was authorised by the Barry Railway in 1893. A smaller dock at 34 acres, it was to be connected to the first dock, now known as the 'Old Dock', by a channel. It opened without ceremony on 10 October 1898. During the period of the building of the first dock at Barry, the Wolfe Barry partnership were undertaking the construction of Tower Bridge (1886–94) for which Henry Marc was responsible for the structural work, including the steel frame that is hidden by the Gothic masonry work of Sir Horace Jones. During the course of the construction of the bridge, Henry Marc sent his pupil Alexander Gibbs on an errand to the site and he narrowly escaped death stepping on a loose plank at a great height.[28]

Sir John Wolfe Barry was the son of the architect Charles Barry, and he shared with Benjamin Baker the position of leading civil engineer of his day. As well as professional engineering works, he was to play a leading role in the establishment of the National Physical Laboratory and the setting up of the Engineering Standards Committee, forerunner of the British Standards Institution. The office was in Delahaye Street, near the Storey's Gate entrance to St James's Park, a site now occupied by the Foreign Office.[29]

This year marks the beginning of modern Barry, the population then being less than 500, and in 1889 the first dock basin was opened, which was to be followed by two additional docks and various port installations. Barry and the surrounding area soon developed in concert with

WOLFE CLOSE
(Named after 'Wolfe Barry,' a Partner of Brunel.)

Wolfe Barry's connection with Barry remembered in a local street name. (SKJ photograph)

Sir John Wolfe Barry (1836–1918), the son of the architect Sir Charles Barry and a pupil of Sir John Hawkshaw, in 1867 started his own engineering practice. Henry Marc became a partner with him in 1878 and together they worked on commissions such as Barry Dock and Tower Bridge. (SKJ collection)

the docks. Many buildings were constructed to cater for the families of the dock construction workers, from houses to small hospitals, which soon went some way toward forming a sizeable town. By 1913, mainly due to its prime location and the railway, Barry had become famous as the largest coal-exporting port in the world, at a time when the docks were crowded with ships and modern ship repair yards, cold stores and flour mills. In 1939 Barry was made into a borough, a significant state of independence for a South Wales town. The Welsh writer and broadcaster Gwyn Thomas (1913–81) referred to the development of Barry, and the new town that the docks created, '… as interesting a stretch of 20th century landscape as we are likely to see in Britain. The town … blew into being as one of the last bursts of volcanic energy that brought Britain through the last century and up to the First World War.'[30]

When Brunel died his passing would be marked in a number of ways. Before the end of 1859 a meeting was held at the offices of Messrs Pritt & Venables in Great George Street, Westminster, which attracted a number of gentlemen, as *The Times* reported: '…most of them members of the profession of which the late Mr. Brunel was so distinguished an ornament … to consider the propriety of erecting a suitable monument to perpetuate his name and the memory of his genius and public services.'

Amongst those present on Saturday 26 November 1859 were fellow engineers Locke, Fowler, Hawkshaw, M'Clean (Robert Stephenson had died on 12 October 1859) and the architect Digby Wyatt. St George Burke QC had organised the meeting and he went on to say that nothing would be more '…graceful and appropriate' than to erect a joint monument to '…perpetuate the memory of Mr. Brunel and Mr. Stephenson'. He went on to say that both were often rivals,

Henry Marc's sketch of Tower Bridge from a letter dated 3 February 1885. (Courtesy of University of Bristol special collections)

but '…still united by the ties of the closest friendship.' But he was sorry to say that there were differences of opinion regarding a joint monument and that it would be '… impossible to carry the project into effect.' Burke was a close friend of Brunel; he had been Parliamentary Agent for the GWR and he would later give material for Isambard's biography of his father, commenting, amongst other things, how he used to accompany Brunel on visits to South Wales to support him at public meetings.[31] The meeting went on to discuss scholarships and the like, but agreed that only a public monument to his memory would suffice. The Recorder of London, Russell Gurney, moved a motion to that effect, which was '…carried with acclamation'. Gurney had addressed the meeting by saying that he was a boyhood friend of Brunel. In time-honoured tradition, a committee was then established which included the engineers Fowler, Hawkshaw and Locke, and the SWR and VNR chairmen C.R.M. Talbot and H.A. Bruce. The result was the statue by Marchetti on the Thames embankment. A stained-glass window in Westminster Abbey would follow later.[32]

A memorial of a much higher profile would be completed in 1864 and would be put into effect by his fellow engineers, particularly Hawkshaw, in the form of the Clifton Suspension Bridge. The bridge was completed using some two-thirds of the suspension links from the demolished Hungerford Suspension Bridge, originally supplied by the Dowlais ironworks.[33] Clifton was therefore completed as both an engineering work and a memorial, reinforcing the sentiment of the famous saying about Wren's works: if you want to see his work look around you. His name would also be fixed on his last major railway work, the Royal Albert Bridge at Saltash. In other areas his sons sought to correct the lack of an engineering biography on their father which had been overlooked by such biographers as Samuel Smiles.[34] This comprehensive, if somewhat impersonal, work was published by Isambard Brunel in 1870.[35] Both were also consulted on a French memorial to their grandfather, Marc Isambard Brunel, who was honoured in his native village of Hacqueville near Gisors in Normandy. The memorial celebrated the centenary of his birth on 25 April 1769.[36] Marc is also commemorated by a London 'blue plaque' at 98 Cheyne Walk and the birthplace of his son is marked by a plaque put up by Portsmouth Council saying: 'In this street was born on 9 April 1806 Isambard Kingdom Brunel The Great Engineer'.[37] Sadly the house it once commemorated was demolished in the 1960s.

There were no monuments put up to the engineer in South Wales; there was the plaque which names him and the Welsh-born chairman of the Bristol & South Wales Union Railway, built to improve communications across the Severn to South Wales – but that was erected on the tunnel portal at Patchway, the railway branch on the Welsh side being part of the SWR. He is remembered in Welsh street names, and ignoring post-industrial examples, you can find examples at Cardiff where Brunel Street can be found near the former GWR Canton locomotive sheds.[38] In the industrial area of Newport, Brunel Street leads to Stephenson Street and the Transporter Bridge. A Great Eastern Terrace can be found appropriately enough in Neyland, overlooking the site of the gridiron that the great ship once rested on, along with a Brunel Avenue. Associations of another kind were the naming of a boat club at Neath as the 'Brunel Boat Club', whose members in 1871 put on dramatic entertainment at the Town Hall, and the Philanthropic Institute's 'Great Eastern Lodge' who met at Ystralyfera in 1863.[39] Brunel's godson, Herbert Isambard Owen (1850–1927), the son of William George Owen, is commemorated on a cornerstone at the University of Wales Registry in Cardiff, and later (in 1950) with a plaque on his birthplace at Chepstow. There were examples of models built to celebrate his works such as the Landore viaduct, of which a scale part section was presented to the Royal Institution of South Wales at Swansea and another model was constructed of his Dare viaduct which was for many years on show at Aberdare Library.

Brunel House as seen from Cardiff Castle Clock Tower.

BRUNEL CLOSE
Named after the famous Engineer who helped design Barry Docks.

For a period after Brunel's death Bristol sought to forget his contribution, in acts such as his ommission from a painting of Bristol worthies. Elsewhere, such as in Northern Ontario, two new townships were established named Brunel and Stephenson.[40] It was in the 1950s that a revival of interest in commemorating Brunel's achievements occurred, and in 1954 two plaques were unveiled, one to commemorate the centenary of Paddington Station and the other a London 'blue plaque' for the *Great Eastern* near the original launch site at Westferry Road on the Isle of Dogs.[41] Interest reached a climax due to the new biography by L.T.C. Rolt, published in 1957 in time for the centenary year of Brunel's death two years later.[42] The interest generated by Rolt's biography is the reason, more than any other, why the first major celebration of the engineer was that year rather than the anniversary of his birth. Plaques were erected at Clifton Suspension Bridge and Saltash Station (to celebrate his Royal Albert Bridge) – but still nothing in South Wales.

Swindon celebrated the development of the new town centre by erecting a copy of the Marchetti Statue, and other statues by Doubleday were put up at Paddington and Bristol. South Wales also saw the first statue of Brunel to be erected in a town he helped to create, Neyland, situated on the marina development that has regenerated the town on the site of the former railway terminus and quay. That development is known, appropriately enough, as Brunel Quay. Henry Marc survived his older brother Isambard, who passed away in 1902, and he is remembered in South Wales with a street name Brunel Close in Barry. A 1980s development, the street name plate records his role in the development of Barry Dock, referring to the 'famous engineer' named Brunel, but not on this particular occassion; Isambard Kingdom Brunel.[43]

Chapter 10 Notes

1 Luxton, Brian C. (first published in 1977, fourth impression 1991), *Old Barry in Photographs* (Stewart Williams: Barry). Foreword by Gwyn Thomas.

2 Barry is as interesting a stretch of twentieth-century landscape as we are likely to see in Britain. The town … blew into being as one of the last bursts of volcanic energy that brought Britain through the last century and up to the First World War. Smith, Dai, Ed. (1986), p.38, *Writer's World; Gwyn Thomas 1913–1981* (Welsh Arts Council: Cardiff).

3 Luxton, Brian C. (First published in 1977, fourth impression 1991).

4 Prothero, Iorwerth W. (1994), p.26.

5 Moore, Donald, Ed. (1984), p.191, Friars Point House is now on the hotel site. The pier (demolished in 1902) was also used by the Yellow Funnel Fleet to pick up and drop visitors to the island and was also known as Traherne's Pier after J.D. Traherne, who bought the house in 1873. For further information on the house see Vale of Glamorgan Treasures (Vale of Glamorgan Council).

6 Guest, Revel and John, Angela V. (1989), p.141, *Lady Charlotte: A Biography of the Nineteenth Century* (Weidenfeld and Nicolson: London).

7 Guest, Revel and John, Angela V. (1989), p.142. The estate had previously been owned by the sister and brother-in-law of John's first wife, Maria Ranken.

8 Vol.I, p.102.

9 Modern-day-Barry consists of three parishes, Cadoxton, Barry and Merthyr Dyfan.

10 This was an attempt by the broad-gauge EVR to promote a narrow-gauge line from Llantrisant to Cardiff, with running powers granted over the EVR from the junction into Llantrisant and the powers to lay narrow-gauge rails over that portion of the EVR. See Vol.II, p.236.

11 Prothero, Iorwerth W. (1994), p.27.

12 Prothero, Iorwerth W. (1994), p.28.

13 Vol.II, Chapter 10, p.236. The EVR was incorporated on 13 June 1857. In fact, the railway had a pre-history linked with promoted and aborted schemes going back beyond the SWR itself, but in 1857 the EVR was promoted as an extension of the SWR from a junction at Llantrisant to Penygraig.

14 Vol.II, Chapter 9, p.236.

15 Guest, Revel and John, Angela V. (1989), p.158.

16 Gwenfo Gynt (2003), p.25, *Wenvoe and Twyn-yr-Odyn* (Tempus Publishing Ltd: Stroud).

17 Jeffreys Jones, T.I. Ed. (1966). p.118, *Acts of Parliament Concerning Wales 1714–1901* (University of Wales Press: Cardiff).

18 Moore, Donald, Ed. (1984), p.213.

19 Jeffreys Jones, T.I. Ed. (1966). pp.190, 123 and 126.

20 Prothero, Iorwerth W. (1994), p.17. From a story in the *Barry Dock News* of 14 March 1893 based on recollections of Mrs Matthews, widow of a former publican of the Wenvoe Arms public house in Wenvoe.

21 This not the public house in Wenvoe but a similarly named public house that existed at the rear of the post office in Vere Street, being moved to its present site in the 1880s with the commencement of Barry Docks and now known as the 'The Admiral'.

22 Moore, Donald, Ed. (1984), p.214. The majority of the £13,000 worth of shares were taken up by Jenner.

23 Price Williams, an engineer related to the coal master Walter Coffin, accompanied Brunel on a number of occasions in South Wales. See Vol.I, p.195.

24 Moore, Donald, Ed. (1984), p.219.

25 Moore, Donald, Ed. (1984), p.220.

26 Robinson, John, *The Barry Dock Works, including the hydraulic machinery and the mode of tipping coal*, Min. Proc. Institution of Civil Engineers, 1889–90, 101, pp.129–151, and *The Barry Graving Docks*, Min. Proc. Institution of Civil Engineers, 1893–94, 116, pp.267–274.

27 Harrison, Godfrey (1950), p.33, *Alexander Gibb, The Story of an Engineer* (Geoffrey Bles: London).

28 Harrison, Godfrey (1950), p.32.

29 Harrison, Godfrey (1950), p.29.

30 Smith, Dai, Ed. (1986), p.38, *Writer's World; Gwyn Thomas 1913–1981* (Welsh Arts Council: Cardiff).

31 Vol.I, p.120.

32 See colour section, Vol.I.

33 Vol.I, pp.89–90.

34 Smiles had written about James Watt, Thomas Telford and both George and Robert Stephenson.

35 Brunel, Isambard (1870, reprinted 1971), *The Life of Isambard Kingdom Brunel* (Longmans, Green & Co.: London, 1870, reprinted by David & Charles: Newton Abbot, 1971).

36 See Vol.I, colour section.

37 See Vol.I, p.24.

38 After Craddock Street (named after a local engineer) it is the first to be reached from Cardiff Central Station followed by Stephenson, Telford and Smeaton Streets with Rennie Street a short distance away.

39 *The Cambrian*, 21 April 1871 and 28 August 1863.

40 Townships in Northern Ontario from at least 1881(?).

41 *The Times*, 31 May 1954.

42 Rolt, L.T.C. (1959), *Isambard Kingdom Brunel* (Longmans, Green & Co.: London).

43 His partner, Wolfe Barry, is also commemorated in the housing development as Wolfe Close.

INDEX